Recent Titles in This Series

(Continued in the back of this publication)

Iterating the Cobar Construction

MEMOIRS
of the
American Mathematical Society

Number 524

Iterating the Cobar Construction

Justin R. Smith

May 1994 • Volume 109 • Number 524 (fourth of 5 numbers) • ISSN 0065-9266

American Mathematical Society
Providence, Rhode Island

1991 *Mathematics Subject Classification.*
Primary 55S45; Secondary 55M99.

Library of Congress Cataloging-in-Publication Data

Smith, Justin R.
    Iterating the cobar construction / Justin R. Smith.
      p. cm. – (Memoirs of the American Mathematical Society, ISSN 0065-9266; no. 524)
    Includes bibliographical references.
    ISBN 0-8218-2588-7
    1. Loop spaces. 2. Cobar construction (Topology) 3. Tensor products. I. Title. II. Series.
QA3.A57 no. 524
[QA612.76]
510 s–dc20   94-4140
[514′.24]   CIP

Memoirs of the American Mathematical Society

This journal is devoted entirely to research in pure and applied mathematics.

Subscription information. The 1994 subscription begins with Number 512 and consists of six mailings, each containing one or more numbers. Subscription prices for 1994 are $353 list, $282 institutional member. A late charge of 10% of the subscription price will be imposed on orders received from nonmembers after January 1 of the subscription year. Subscribers outside the United States and India must pay a postage surcharge of $25; subscribers in India must pay a postage surcharge of $43. Expedited delivery to destinations in North America $30; elsewhere $92. Each number may be ordered separately; *please specify number* when ordering an individual number. For prices and titles of recently released numbers, see the New Publications sections of the *Notices of the American Mathematical Society.*

    Back number information. For back issues see the *AMS Catalog of Publications.*

    Subscriptions and orders should be addressed to the American Mathematical Society, P. O. Box 5904, Boston, MA 02206-5904. *All orders must be accompanied by payment.* Other correspondence should be addressed to Box 6248, Providence, RI 02940-6248.

Memoirs of the American Mathematical Society is published bimonthly (each volume consisting usually of more than one number) by the American Mathematical Society at 201 Charles Street, Providence, RI 02904-2213. Second-class postage paid at Providence, Rhode Island. Postmaster: Send address changes to Memoirs, American Mathematical Society, P. O. Box 6248, Providence, RI 02940-6248.

10 9 8 7 6 5 4 3 2 1   99 98 97 96 95 94

Contents

ABSTRACT. This paper develops a new invariant of a CW-complex called the *m-structure* and uses it to perform homotopy-theoretic computations. The m-structure of a space encapsulates the coproduct structure, as well as higher-coproduct structures that determine Steenrod-operations. Algebraically, it amounts to an operad in the category of modules. In particular, given an m-structure on the chain complex of a reduced simplicial complex of a pointed simply-connected space, one can equip the cobar construction of this chain-complex with an natural m-structure. The m-structure of the cobar construction is shown to be geometrically meaningful, in the sense that it corresponds to the m-structure of the loop space of the original space under the map that carries the cobar construction to the loop space.

This result allows one to form *iterated* cobar constructions that are shown to be homotopy equivalent to iterated loop-spaces. This homotopy equivalence is in the sense of chain-complexes equipped with m-structures.

These results are applied to the computation of the cohomology algebra structure of total spaces of fibrations (actually, we compute m-structures, which determine the cohomology algebra).

Key words and phrases. coproduct, cobar construction, twisted tensor products, cohomology operations.

CHAPTER 1

Introduction

One of the most important invariants of homotopy type of a topological space is the coproduct-structure on the chain-complex. Indeed, it determines the rational homotopy type of a pointed simply-connected space (see [20]). Over the integers there are many additional invariants of homotopy type including Steenrod operations on the mod p cohomology rings for all primes p. In this paper we will present an algebraic theory that incorporates all of these invariants, and allows one to compute them for:

- the loop-space of a space (via the cobar construction) and;
- the total space of a fibration;

A key element of computing the coproduct of the total space of a fibration is the determination of the coproduct on the chain complex of the loop space of the base. Since Adams showed that this chain-complex is given by the *cobar construction* (see [1]), we would like to know a geometrically induced coproduct on the cobar construction. Here, the term "geometrically induced coproduct" can be defined in several ways with varying degrees of strength. We essentially regard the Alexander-Whitney diagonal map on a simplicial chain-complex as being the canonical geometric one and any other diagonal homotopic to it as being geometric to some extent.

In the rational case Quillen showed (in [20]) that the shuffle coproduct on the cobar construction is geometric, where this is a dual of the shuffle product defined by Eilenberg and MacLane for the bar construction in [5]. This result implied a number of other results that made it relatively easy to compute a geometric coproduct on the total space of a fibration and on chain-complexes of simply-connected pointed spaces in general. In the integral case the shuffle coproduct on the cobar construction remains well-defined, in some cases, but Quillen's proof of its geometricity is no longer valid. In fact, any attempt to find a geometrically valid coproduct on the cobar construction encounters the following two obstacles, discovered by Alain Prouté:

> The shuffle coproduct on the integral cobar construction is demonstrably non-geometric — see [19]. Here the term 'geometric' is defined in a very weak sense — the shuffle coproduct is non-geometric to the extent that it even induces the wrong maps in homology.

Received by the editor October 14, 1992.

1

The first person to obtain a geometric coproduct on the cobar construction was Hans Baues, in [2]. Baues' approach to this problem was to find a cellular description of the loop-space analogous to Milgram's cellular bar construction (in [16]. The chain-complex of this cellular loop-space coincides with Adams' cobar construction.

When a space is a simplicial set the cells in Baues' cobar construction correspond to tuples of simplices in the original space. In this case Baues gave a formula for the coproduct of a cell represented by a single simplex of the original space. The formula involved *face operations* on the simplex that corresponded to the cell. It was then extended to the entire cobar construction, and defined a co-commutative, geometric, coproduct on the cobar construction.

Baues was able to use this coproduct to iterate the cobar construction once. Unfortunately, his geometric methods cannot, apparently, be extended to give a coproduct on *this* iterated cobar construction. Since his formula makes *explicit* use of the simplicial structure of the original space, one cannot iterate it without a natural simplicial decomposition of the cells that occur in his cobar construction.

In § 4 of chapter 2 (page 29) we derive a canonical m-structure, denoted $\mathcal{C}(X)$, for the chain-complex of a simplicial complex, X. This is essentially a chain-complex, $\mathcal{C}(X)$, equipped with S_n-equivariant mappings $\tilde{f}_n\colon \mathrm{R}(S_n) \otimes \mathcal{C}(X) \to (\mathcal{C}(X))^n$, for all $n \geq 1$, where S_n is the symmetric group, $\mathrm{R}(S_n)$ is the bar-resolution and S_n acts on $\mathcal{C}(X)^n$ by permutation of factors. The construction of $\mathcal{C}(X)$ proceeds by an argument that essentially boils down to acyclic models. The lowest-degree component of the m-structure of $\mathcal{C}(X)$, i.e. $\tilde{f}_n|[\,] \otimes \mathcal{C}(X)$, coincides with the Alexander-Whitney diagonal map (iterated n times), hence may be regarded as determining the cup-product structure of cohomology. Higher components of this m-structure determine all Steenrod operations (see 3.4 on page 23). Some of the motivation for our definition of m-coalgebras came from a paper of James Davis ([6]).

Coherency turns out to be a key feature of the m-structures used in this paper. Aside from the obvious implications (i.e. associativity of the underlying coproduct) it implies that compositions of higher coproducts are canonically equal to certain other higher coproducts — see remark 3.3.1 on page 21. For instance, let Δ_a denote $\mathfrak{D}_n(\alpha \otimes *)$. Then it turns out that: $\Delta_{[(1,2)]} \otimes 1 \circ \Delta_{[(1,2)]} = \Delta_{[(1,3,2)|(1,2)]} - \Delta_{[(1,2)|(1,2,3)]}$. We also derive coherent m-structures on the bar construction of a DGA-algebra that is equipped with a coherent m-structure. Coherency essentially provides us with a "dictionary" that we can use to convert formulas involving compositions of higher coproducts into formulas involving elements of the $\{\mathrm{R}(S_n)\}$.

The basic definitions in chapter 2 can be stated in terms of operads in the category of graded differential modules. Operads were originally defined in terms of topological spaces by May in [15] and this concept was extended to DG-modules by Smirnov in [22]. Essentially:

- the set of bar resolutions (over $\mathbb{Z}$) $\{\mathrm{R}(S_n)\}$ of symmetric groups, equipped with the operation defined in 2.17 on page 17, constitutes an operad, and
- the functor $\mathcal{C}(X)$ is a coalgebra over this operad, in the sense of § 2 of [22].

In § 2 in chapter 3 (page 38) we consider the question of the existence of m-structures on the cobar construction of a chain-complex, C, that is itself equipped with a weakly coherent m-structure. We determine a family of such m-structures on the cobar construction, $\mathcal{F}(C)$, of C and show that:

- they are all natural with respect to m-structure homomorphisms of C;
- they are all naturally homotopic to one another.

The basic idea is as follows:

1. Suppose that C is equipped with a coherent m-structure.
2. Now assume, roughly speaking, that $\mathcal{F}(C)$ (the cobar construction) has a coproduct that can be expressed in terms of the m-structure of C. In other words simply write down the coproduct of $\mathcal{F}(C)$ restricted to $\Sigma^{-1}C \subset \mathcal{F}(C)$ and regard it as a map $f\colon C \to C^k$ (after suspending, suitably). Now regard that map as $\mathfrak{D}(\sigma \otimes *)$ for some $\sigma \in \mathrm{R}(S_k)$. We can use the coherency of the m-structure of C (and the "dictionary" mentioned above) to convert the obvious conditions on f (imposed by the way the differential of $\mathcal{F}(C)$ is defined) into conditions on the element $\sigma \in \mathrm{R}(S_k)$. Now solve for elements of $\mathrm{R}(S_k)$ that satisfy these algebraic conditions, and we have a coproduct on $\mathcal{F}(C)$.

This discussion is somewhat over-simplified but contains the main idea of § 2 of chapter 3 (page 38). The main result of this section is:

Theorem 2.30 (from page 57): *Let C be a weakly-coherent m-coalgebra, with $C_0 = \mathbb{Z}$ and $C_i = 0$ for $0 < i < k$, where $k \geq 2$. Then $\mathcal{F}(C)$ has a natural structure as a strict weakly-coherent m-Hopf-algebra.*

§ 3 of that chapter proves similar results for the bar construction of an m-coalgebra. The m-structure on the bar construction is proved to be geometrically valid.

In chapter 3 in § 4 the results of §§ 2 and 3 are applied to prove the geometric validity of the m-structure on the cobar construction. The main result is derived from 2.27 on page 55 and 4.2 on page 66 — it amounts to an explicit formula for computing this m-structure:

Theorem: *Let X be a pointed simplicial complex with loop space ΩX and suppose the chain-complexes of X and ΩX are equipped with the coherent m-structures defined on page 29. Let the adjoint maps to the structure maps be $\tilde{f}_n\colon \mathrm{R}(S_n) \otimes \mathcal{C}(X) \to \mathcal{C}(X)^n$. Then there exists a natural chain-homotopy equivalence of DGA-algebras $\Phi\colon \mathcal{F}(\mathcal{C}(X)) \to \mathcal{C}(\Omega X)$ that fits into chain homotopy-commutative diagram:*

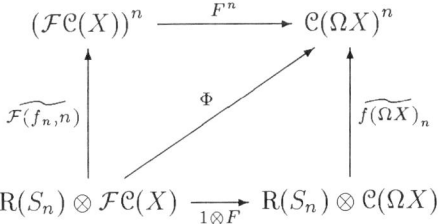

for all n, where

1.

$$\widetilde{\mathcal{F}(f_n, n)} = \sum_{\alpha_1, \ldots, \alpha_n} \underbrace{\downarrow \otimes \cdots \otimes \downarrow}_{|\alpha| \text{ factors}} \circ \tilde{f}_n \circ (z_{\alpha_1, \ldots, \alpha_n} \otimes \uparrow) \colon \mathrm{R}(S_n) \otimes \Sigma^{-1}\mathcal{C}(X)^+$$
$$\to \bigoplus_{\alpha_1, \ldots, \alpha_n} \left(\Sigma^{-1}\mathcal{C}(X)\right)^{\alpha_1} \otimes \cdots \otimes \left(\Sigma^{-1}\mathcal{C}(X)\right)^{\alpha_n} \subset \mathcal{F}(\mathcal{C}(X))^n$$

where the $\{z_{\alpha_1,\ldots,\alpha_n}\}$ are functions $z_{\alpha_1,\ldots,\alpha_n}\colon R(S_n) \to R(S_{|\alpha|})$ (of degree $|\alpha| - 1$), and the summation is taken over all sequences $\alpha_1,\ldots,\alpha_n \geq 0$, $|\alpha| = \sum_{i=1}^n \alpha_i$. This definition is extended to all of $\mathcal{F}(\mathcal{C}(X))$ by defining it to be a DGA-algebra morphism.

2. *$f_{\mathcal{C}(\Omega X)}$ is the canonical m-structure that is associated with the functor $\mathcal{C}(\Omega X)$. In addition the chain-homotopy, ϕ, is natural with respect to simplicial maps of X.*

Remarks.

1. The maps $\uparrow\colon \Sigma^{-1}C \to C$ and $\downarrow\colon C \to \Sigma^{-1}C$ are, respectively, the suspension and desuspension isomorphisms. Recall that the chain-complex of $\mathcal{F}(C)$ is the tensor algebra over $\Sigma^{-1}C^+$ — see 2.1 on page 38.

2. The unpleasant subscripts under the z's are due to the fact that we cannot regard the coproduct on $\mathcal{F}(C)$ as just a map $f\colon C \to C^k$ — its target is really a direct sum of C^i for various values of i.

3. Although the sum over $\{\alpha_1,\ldots,\alpha_n\}$, as written, is infinite, all but a finite number of terms vanish when it is evaluated on any element of $R(S_n) \otimes \Sigma^{-1}\mathcal{C}(X)^+$. This is due to the fact that the image of the maps lies in a product of $|\alpha|$ copies of $\Sigma^{-1}\mathcal{C}(X)^+$, and $= 0$, and it is the adjoint of a map $\mathcal{C}(X) \to \mathrm{Hom}_{\mathbb{Z}S_n}(*, *)$.

An algorithm for computing the functions $\{z_{\alpha_1,\ldots,\alpha_n}\}$ is given in 2.27 on page 55 and the surrounding discussion. It amounts to a recursive formula for z_α in terms of $z_{\alpha'}$, where the sequence α' is lexicographically less than α. In the case where the original m-coalgebra was coherent (rather than weakly coherent) this formula is somewhat simpler than in the general case — it is:

$$\partial \circ z_\alpha = (-1)^{|\alpha|-1} z_\alpha \circ \partial + (-1)^{|\alpha|} \sum_{\{\alpha'+\alpha''=\alpha\}} (-1)^{(|\alpha'|-1)|\alpha''|}(z_{\alpha'} \cup z_{\alpha''})$$

$$+ (-1)^{|\alpha|} \sum_{j=1}^{|\alpha|-1} (-1)^j \mathfrak{T}_j \circ z_{(\alpha,j)}$$

where $\alpha = (\alpha_1,\ldots,\alpha_n)$, $\alpha' = (\alpha'_1,\ldots,\alpha'_n)$, $\alpha'' = (\alpha''_1,\ldots,\alpha''_n)$ and the notation $\{\alpha' + \alpha'' = \alpha\}$ denotes componentwise addition. Here the cup-product in the term $(z_{\alpha'} \cup z_{\alpha''})$ is taken using the canonical coproduct on the bar resolutions $\{R(S_n)\}$ and a product operation[1] $\cdot\colon R(S_{\alpha'}) \otimes R(S_{\alpha''}) \to R(S_\alpha)$ (here $R(S_{\alpha'})) = R(S_{|\alpha'|})$) that is the (twisted) shuffle product times the signed permutation that shuffles the two sequences $\alpha' = (\alpha'_1,\ldots,\alpha'_n)$, $\alpha'' = (\alpha''_1,\ldots,\alpha''_n)$ together. See 2.27 on page 55 for the general result.

Some computations are given in appendix F (page 119) for the case where there are *two* subscripts. Note that we have computed a geometric coproduct on $\mathcal{F}(C)$ in the following sense: we have computed an m-structure on $\mathcal{F}(C)$ homotopic to the canonical one on ΩX, and the lowest-dimensional component of that m-structure is the Alexander-Whitney diagonal map. The higher components of this m-structure determine Steenrod operations of all ranks and degrees.

The computations in appendix F, for instance, imply the following statement:

---

[1] Although $R(S_{\alpha'}) = R(S_{|\alpha'|})$, we write the subscript as a sequence because the product depends upon the terms of this sequence.

If $C_i = 0$ for $i \neq 0$ and $i < k$ then there exists an isomorphism $\downarrow^*\colon H^i(\mathcal{F}(C); \mathbb{Z}_2) \to H^{i+1}(C; \mathbb{Z}_2)$ in the stable range $(i < 2k)^2$. Our formula for the coproduct on $\mathcal{F}(C)$ implies that if $y \in H^i(\mathcal{F}(C); \mathbb{Z}_2)$ then $\downarrow^*(y \cup y) = Sq^{i-1}\downarrow^*(y)$, $i < 2k$. This shows that (at least mod 2) squares in the cobar construction map into certain Steenrod squares in the original chain-complex. Unfortunately, the coproduct in general (or the cup product in cohomology) on the cobar construction, doesn't appear to be as easily expressible in terms of commonly used cohomology operations. The reduction mod 2 kills off the bilinear data that would be needed.

These results can be used to iterate the cobar construction and it is proved that the resulting induced coproduct is also geometric:

Corollary 4.3 (page 67): *Let X be an n-connected pointed simplicial complex equipped with the canonical (coherent) m-structure, $\mathcal{C}(X)$. If $i < n$ is an integer then there exists a morphism of m-coalgebras $\mathcal{F}^{i-1}(u)\colon \mathcal{F}^i\mathcal{C}(X) \to \mathcal{C}(\Omega^i X)$, that induces a homotopy equivalence of underlying chain-complexes and is a morphism of DGA-algebras.*

I feel it appropriate at this time to compare the results of this paper with some of those of Smirnov (in his remarkable paper [23]) and several of his other papers. Our definition of coherent m-structure resembles Smirnov's concept of canonical coalgebra structure over the operad E^* (defined in [23]) in the following sense:

Smirnov's canonical coalgebra structure over E^* is equivalent to a family of maps $\hat{S}_n\colon \mathcal{C}(X) \to \hom(E(n), (\mathcal{C}(X))^n)$ that satisfy a coherency condition very similar to ours. This Smirnov structure expresses much of the same information as a coherent m-structure — in fact Smirnov proves (in § 6 of [23]) that his structures determine the weak homotopy type of the space X. The differences between Smirnov's invariant and ours may be summed up thusly:

1. Smirnov's structure is more complex than ours: the chain-complexes $E(n)$ are uncountably generated in all dimensions and for all values of $n > 1$. Although there is no obvious connection between Smirnov's invariant and other more commonly used homotopy invariants (such as Steenrod operations, coproducts, etc.), he develops a connection with Steenrod squares in § 4 of [23]) and other structures of the Steenrod algebra in [30]. In [31], Smirnov shows that the chain-complex of E^* has the structure of an *f-resolution* as we define it in 3.1 on page 19 — although Smirnov doesn't use this terminology.

2. Smirnov's invariant may be more powerful than ours — although we have (unpublished) results that show that our invariant determines the weak-homotopy type of simply-connected, pointed space, these restrictions don't exist for Smirnov's invariant. In addition, Smirnov's invariant solves the question of topological realizability of chain-complexes in the sense that a chain-complex is topologically realizable if and only if it admits a Smirnov structure.

3. Smirnov defines a form of the cobar construction for chain-complexes equipped with a Smirnov structure (see § 3 of [23]) but it does not appear to lend itself to computations to the extent that the conventional cobar construction used in the present paper (roughly speaking, it is to the

---

[2] This is the connecting map in the Serre exact sequence of the canonical acyclic twisted tensor product $X \otimes_a \mathcal{F}(C)$.

conventional cobar construction as the W-construction is to the bar construction — see [5]). Smirnov's and the conventional bar construction are, of course, homotopy equivalent in a canonical sense.

Chapter 4 on page 69 applies these results to the computation of m-structures on the chain-complex of the total space of a fibration.

The *chain-complex* of the total space of a fibration has been computed by a number of authors (and a number of different points of view), beginning with Brown, in [3], R. H. Szczarba in [28], Shih Weishu in [32], and V. K. A. M. Gugenheim in [10].

The rational *cohomology algebra* of the total space has also been computed from several points of view:

- as a by-product of Quillen's work on rational homotopy theory — see [20].
- Using the Lie algebra model of homotopy types, by Tanré— see [29].
- using the filtered model of homotopy types defined by Halperin and Stasheff in [12]. See Saneblidze's paper, [21].

In [17], Proutéexplored the question of how one computes the mod-p cohomology algebra of the total space of a fibration.

Our strategy for computing the coproduct (and the m-structure) of the integral chain-complex of the total space of a fibration involves a sequence of steps.

1. Given a weakly-coherent m-coalgebra, C, we compute a functorial weakly-coherent m-structure on $C \otimes_\ell \mathcal{F}(C)$. This twisted tensor product is the canonical acyclic tensor-product with base equal to C — it is defined in 2.3 on page 39.

2. We show that the geometric m-structure on an arbitrary twisted tensor product $C \otimes_\xi F$ can be regarded as being induced by that of $C \otimes_\ell \mathcal{F}(C)$. If A is a DGA coalgebra, $\xi\colon A \to F$ is a twisting cochain, and Z is a right A-comodule then $Z \otimes^A (A \otimes_\xi F) = Z \otimes_\xi F$, where $* \otimes^A *$ is the *cotensor product*, defined in 1.17 on page 84.

   We, consequently, find a formula for the m-structure of the cotensor product of a comodule and a coalgebra. This is done in 1.19 on page 84, and in § 2 beginning on page 86.

3. In § 3 on page 93 we show that the m-structure that we get on a twisted tensor product is *geometric*. It is geometric in the sense that it is homotopic to the m-structure derived from $\mathcal{C}(*)$ (which is essentially derived from *acyclic models*). This essentially follows from the acyclicity of $C \otimes_\ell \mathcal{F}(C)$ — which implies that *all* m-structures on $C \otimes_\ell \mathcal{F}(C)$ are homotopic. It turns out that we can transport the homotopies that occur in this fashion to *cotensor products*. This is done in 3.2 on page 94, and 3.3 on page 94.

In order to carry out the first step listed above, we must find a functorial m-structure on $C \otimes_\ell \mathcal{F}(C)$. This is similar to, but more difficult than, the computation of a canonical m-structure on $\mathcal{F}(C)$, outlined above.

We define an m-structure on $C \otimes_\ell \mathcal{F}(C)$ by giving a sequence of maps (in 1.5 on page 72):

$$\mathcal{A}_n\colon C \to \mathrm{Hom}_{\mathbb{Z}S_n}(\mathcal{P}(C,n), (C \otimes_\ell \mathcal{F}(C))^n)$$

where $\mathcal{P}(C,n)$ is a $\mathbb{Z}S_n$-resolution of $\mathbb{Z}$ defined in 2.5.1 on page 92. This is a type of twisted tensor product $\mathcal{P}(\mathfrak{R}[C], n) = \hat{Y}_n(\mathfrak{R}[C]) \bigstar_{\{\ell, \hat{\mu}_i\}} \mathcal{F}(\mathfrak{R}[C], n)$, where $\hat{Y}_n(\mathfrak{R}[C])$

is the inverse limit of the inverse system $q_{m',m} \colon Y_{n,m'}(\mathfrak{R}[C]) \to Y_{n,m}(\mathfrak{R}[C])$, defined in 1.1 on page 69 and the coordinate coalgebra, $\mathcal{F}(\mathfrak{R}[C], n)$, of $\mathcal{F}(C)$. The type of twisted tensor product used here is defined in 2.5 on page 40.

For any sequence, α, of length n of nonnegative integers let $y_\alpha \colon R(S_n) \to \mathfrak{R}[C]_{|\alpha|+n}$ be maps of degree $|\alpha|$ and define $c_n(\alpha) = \downarrow^{|\alpha|} \circ y_\alpha \colon R(S_n) \to Y_{n,m}(\mathfrak{R}, \alpha)$ and $c_{n,m} \colon R(S_n) \to Y_{n,m}(\mathfrak{R})$ to be

$$\sum_{\substack{\alpha \\ |\alpha| \le m}} c_n(\alpha)$$

See page 81 for the definitions of these maps.

We can evaluate the m-structure on $C \otimes_\ell \mathcal{F}(C)$ by finding chain-maps $u_{n,m} = c_{n,m} \otimes (p_m \circ h_n) \circ \Delta_R \colon R(S_n) \to \mathcal{P}(R(S_n), n)$, and composing them with the maps $\mathcal{A}_n$. The requirement that the m-structure of $C \otimes_\ell \mathcal{F}(C)$ project to that of C implies that $y_0 =$ the structure map of C, where 0 in this context represents the sequence of n zeroes.

THEOREM: (1.14 on page 82) *The* $\{y_\alpha\}$, *defined above must satisfy the following formula:*

$$y_\alpha \circ \partial_R - (-1)^{|\alpha|} \partial_{\mathfrak{R}} \circ y_\alpha$$

$$- \sum_{t=1}^{|\alpha|} \sum_{i=2}^{\infty} (-1)^{|\alpha| i + (i+1)(t+n)} \Delta_i \circ_{t+n} y_\alpha(t+n, i) +$$

$$\sum_{i=2}^{\infty} \sum_{t=2}^{n} (-1)^{|\alpha| i} \mathcal{Z}\Big\{ (t, i-1), \big(n - t + \sum_{j=1}^{t-1} \alpha_j, \sum_{j=t}^{n} \alpha_j \big) \Big\} \Delta_i \circ_t y_\alpha(t, i)$$

$$- \sum_{i=2}^{\infty} (-1)^{i|\alpha|} \sum_{\beta_1 + \cdots + \beta_i = \alpha} (-1)^{\sum_{j=2}^{i} |\beta_j| \left(|\beta_1| + \sum_{j=2}^{i-1} |\beta_j| - 1 \right)} \mathcal{S}_n(\mathcal{Z}\{\beta_1, \ldots, \beta_i\})$$

$$\circ t_i \circ y_{\beta_1} \otimes z_{\beta_2} \otimes \cdots \otimes z_{\beta_i} \circ \Delta_{\mathcal{F}}^{i-1} = 0$$

Remarks.

1. This formula allows us to calculate the $\{y_\alpha\}$, given contracting chain-homotopies for the $Y_{n,m}(\mathfrak{R})$ — it is particularly suited to machine-computation. It expresses the boundary of a given y_α in terms of the $\{z_\alpha\}$ and other $y_{\alpha'}$, where α' is lexicographically less than α. The fact that $Y_{n,m}(\mathfrak{R})$ is acyclic in positive dimensions implies that we can do an inductive computation of the $\{y_\alpha\}$.

2. The $y_\alpha(u, v)$ are defined as follows (compare remark 2.27.2 on page 55): Let $\gamma(i)$ be defined to be the *largest* value of t such that $\sum_{k=1}^{t-1} \alpha_k < i \le \sum_{k=1}^{t} \alpha_k$.
   Then $y_\alpha(i, j)$ is defined to be $\begin{cases} 0 & \text{if } \alpha_{\gamma(i+j-1)} \le j - 1 \\ y_{\alpha'} & \text{if } \alpha_{\gamma(i+j-1)} > j - 1 \end{cases}$, where α' is the
   sequence that results from subtracting $j - 1$ from the $\gamma(i + j - 1)^{\text{th}}$ term of α.

3. The $\{t_i\}$ are defined in remark 2.14.2 on page 45.

Given an m-structure on $C \otimes_\ell \mathcal{F}(C)$, we can compute an m-structure on a twisted tensor product that represents the total space of a fibration. This is done in § 2 on page 86 and § 3 on page 93.

One of the main results of that section is:

COROLLARY (3.5 on page 96) *Let $f\colon X \to Y$ be a map of pointed simply connected spaces. The homotopy-fiber of this map is homotopy-equivalent (as an m-coalgebra) to the weakly-coherent m-coalgebra $(\mathcal{C}(X)_\ell \mathcal{F}(\mathcal{C}(X))) \otimes_{\mathcal{F}(\mathcal{C}(X))} \mathcal{F}(\mathcal{C}(Y)) = \mathcal{C}(X) \otimes_{\mathcal{F}(f) \circ \ell} \mathcal{F}(\mathcal{C}(Y))$.*

The chain-maps $u_{n,m} = c_{n,m} \otimes (p_m \circ h_n) \circ \Delta_R \colon \mathrm{R}(S_n) \to \hat{Y}_n(\mathfrak{R}) \bigstar_{\{\ell, \hat{\mu}_i\}} \mathcal{F}(\mathfrak{R}, n)$, mentioned above, can be used to explicitly compute the m-structure on twisted tensor products. We form the composite of the structure maps

$$\mathfrak{f}[\mathcal{C}(X) \otimes_{\mathcal{F}(f) \circ \ell} \mathcal{F}(\mathcal{C}(Y))]_n \colon \mathcal{C}(X) \otimes_{\mathcal{F}(f) \circ \ell} \mathcal{F}(\mathcal{C}(Y))$$
$$\to \mathrm{Hom}_{\mathbb{Z} S_n}(\hat{Y}_n(\mathfrak{R}) \bigstar_{\{\ell, \hat{\mu}_i\}} \mathcal{F}(\mathfrak{R}, n), (\mathcal{C}(X) \otimes_{\mathcal{F}(f) \circ \ell} \mathcal{F}(\mathcal{C}(Y)))^n)$$
$$\to \mathrm{Hom}_{\mathbb{Z} S_n}(\mathrm{R}(S_n), (\mathcal{C}(X) \otimes_{\mathcal{F}(f) \circ \ell} \mathcal{F}(\mathcal{C}(Y)))^n)$$

where the map on the right is $\mathrm{Hom}_{\mathbb{Z} S_n}(u_n, 1)$. This composite can be used to compute higher-coproducts for $\mathcal{C}(X) \otimes_{\mathcal{F}(f) \circ \ell} \mathcal{F}(\mathcal{C}(Y))$.

I would like to thank James Stasheff for his considerable encouragement in completing this project. I am also indebted to Hans Baues for several interesting and informative discussions, and many valuable suggestions. These suggestions resulted in notable improvements to the exposition. I am, of course, entirely responsible for any flaws or errors in this paper.

CHAPTER 2

m-coalgebras

1. Preliminaries

In this section we define m-structures and the coherency of such structures.

DEFINITION 1.1. Let C and D be two graded $\mathbb{Z}$-modules. A map of graded modules $f\colon C_i \to D_{i+k}$ will be said to be of degree k.

REMARK. 1.1.1. For instance the *differential* of a chain-complex will be regarded as a degree -1 map.

We will make extensive use of the Koszul Convention (see [9]) regarding signs in homological calculations:

DEFINITION 1.2. If $f\colon C_1 \to D_1, g\colon C_2 \to D_2$ are maps, and $a \otimes b \in C_1 \otimes C_2$ (where a is a homogeneous element), then $(f \otimes g)(a \otimes b)$ is defined to be $(-1)^{\deg(g)\cdot\deg(a)} f(a) \otimes g(b)$.

Remarks. 1.2.1. This convention simplifies many of the common expressions that occur in homological algebra in particular it eliminates complicated signs that occur in these expressions. For instance the differential, $\partial_\otimes$, of the tensor product $C \otimes D$ is just $\partial_C \otimes 1 + 1 \otimes \partial_D$.

1.2.2. Throughout this entire paper we will follow the convention that group-elements act on the *left*. Multiplication of elements of symmetric groups will be carried out accordingly i.e. $(1, 2, 3) * (1, 4)$ = result of applying $(1, 4)$ first and then $(1, 2, 3) = (1, 4, 2, 3)$ rather than $(1, 2, 3, 4)$.

1.2.3. Let $f_i, g_i, i = 1, 2$, be maps. It isn't hard to verify that the Koszul convention implies that $(f_1 \otimes g_1) \circ (f_2 \otimes g_2) = (-1)^{\deg(g_1)\deg(f_2)}(f_1 \circ f_2 \otimes g_1 \circ g_2)$.

1.2.4. We will also follow the convention that, if f is a map between chain-complexes, $\partial f = \partial \circ f - (-1)^{\deg(f)} f \circ \partial$. The *compositions* of a map with boundary operations will be denoted by $\partial \circ f$ and $f \circ \partial$ – see [9]. This convention clearly implies that $\partial(f \circ g) = (\partial f) \circ g + (-1)^{\deg(f)} f \circ \partial g$. We will call any map f with $\partial f = 0$ a *chain-map*. We will also follow the convention that if C is a chain-complex and $\uparrow\colon C \to \Sigma C$ and $\downarrow\colon C \to \Sigma^{-1}C$ are, respectively, the *suspension* and *desuspension* maps, then $\uparrow$ and

9

$\downarrow$ are both *chain-maps*. This implies that the boundary of ΣC is $- \uparrow \circ \partial_C \circ \downarrow$ and the boundary of $\Sigma^{-1}C$ is $- \downarrow \circ \partial_C \circ \uparrow$.

1.2.5. We will use the symbol T to denote the *transposition operator* for tensor products of chain-complexes $T: C \otimes D \to D \otimes C$, where $T(c \otimes d) = (-1)^{\dim(c)\dim(d)}d \otimes c$, and $c \in C, d \in D$.

2. Formal coalgebras

In this section we will develop an algebraic construct called a formal coalgebra. Formal coalgebras model some of the formal properties of coalgebras and facilitate the definition of coherent m-coalgebras. In order to motivate the concept, we will begin with an example:

Let C be a DGA-module with augmentation $\epsilon: C \to \mathbb{Z}$, and with the property that $C_0 = \mathbb{Z}$. We want to model "coalgebra structures" on C. This immediately brings to mind a map $\Delta: C \to C^2$, where $C^2 = C \otimes_{\mathbb{Z}} C$. We will be interested in coalgebra structures that are not necessarily co-associative. This means that we must consider composites $\{\Delta, (1 \otimes \Delta) \circ \Delta, (\Delta \otimes 1) \circ \Delta, \dots\}$. It follows that we must be concerned with maps $C \to C^n$ for all $0 \le n < \infty$. Here $C^n = \underbrace{C \otimes \cdots \otimes C}_{n \text{ factors}}$ and by convention,

we will regard $C^0 = \mathbb{Z}$, concentrated in dimension 0.

It follows, that if we want to describe the set of generalized coalgebra structures on the chain-complex C, we must consider:

- $\mathrm{Hom}_{\mathbb{Z}}(C, C^n)$ for all $0 \le n < \infty$. Note that each element in $\mathrm{Hom}_{\mathbb{Z}}(C, C^n)$ has two numbers associated with it: its *rank*, which simply records the fact that it is a map to C^n (i.e., it has n copies of C in its target), and its *dimension*, which is equal to its *degree* as a map of chain-complexes — see 1.1 on page 9.
- given $\mathrm{Hom}_{\mathbb{Z}}(C, C^n)$ and $\mathrm{Hom}_{\mathbb{Z}}(C, C^m)$, we must consider the ways of composing maps contained in these modules. Composition defines bilinear pairings $\circ_i: \mathrm{Hom}_{\mathbb{Z}}(C, C^n) \otimes \mathrm{Hom}_{\mathbb{Z}}(C, C^m) \to \mathrm{Hom}_{\mathbb{Z}}(C, C^{n+m-1})$, where $\circ_i$ represents the composition of maps that plugs a map of $\mathrm{Hom}_{\mathbb{Z}}(C, C^n)$ into the ith factor of the target of a map of $\mathrm{Hom}_{\mathbb{Z}}(C, C^m)$. In other words, if $f \in \mathrm{Hom}_{\mathbb{Z}}(C, C^m)$ and $g \in \mathrm{Hom}_{\mathbb{Z}}(C, C^n)$, $f \circ_i g$ is the composition:

$$
\begin{array}{ccc}
C & \xrightarrow{\quad f \quad} & C^m = C^{i-1} \otimes C \otimes C^{m-i} \\
& & \Big\downarrow \underbrace{1 \otimes \cdots 1 \otimes g \otimes 1 \otimes \cdots 1}_{i^{\text{th}} \text{ position}} \\
& C^{i-1} \otimes C^n \otimes C^{m-i} \xrightarrow{\quad = \quad} & C^{n+m-1}
\end{array}
$$

Clearly, i can only run from 1 to m.

Careful consideration of how the compositions are defined implies that these operations satisfy the following "associativity" conditions:

- $(u \circ_i v) \circ_j w = u \circ_{i+j-1} (v \circ_j w)$;
- if $j < i$ then $u \circ_{i+\mathrm{rank}(v)-1} (v \circ_j w) = (-1)^{\deg(u)\cdot\deg(v)}v \circ_j (u \circ_i w)$ (by the Koszul convention).

With this in mind, we are ready to define formal coalgebras:

DEFINITION 2.1. Let $U = \{U_n\}$ denote a sequence of differential graded $\mathbb{Z}$-free chain-complexes with n running from 0 to ∞. This sequence will be said to constitute a *formal coalgebra* with U_n being the component of rank n if:

Given nonnegative integers n, m with $n + m \geq 1$ the following bilinear pairings are defined:

$$\circ_i \colon U_n \otimes U_m \to U_{m+n-1}$$

where $1 \leq k \leq m$ and the pairings respect the gradings (in other words $a \in U_n, b \in U_m$ implies that $\dim(a \circ_i b) = \dim(a) + \dim(b)$).

The composition-operations are subject to the following identities Given nonnegative integers n, m, t, and elements, $a \in U_n, b \in U_m$ and $c \in U_t$

Associativity Identity $(a \circ_i b) \circ_j c = a \circ_{i+j-1} (b \circ_j c)$;

Commutativity Identity if $j < i$ then

$$a \circ_{i+\text{rank}(b)-1} (b \circ_j c) = (-1)^{m \cdot n} b \circ_j (a \circ_i c)$$

The differential $\partial \colon U \to U$:

1. preserves rank;
2. imposes the following additional condition on composition operations $\partial(a \circ_i b) = \partial a \circ_i b + (-1)^{\dim(a)} a \circ_i \partial b$.

Remarks. 2.1.1. The module $\bigoplus_{k=0}^{\infty} U_k$ has a $\mathbb{Z}$-basis of elements b with the property that each basis element has *two* numbers associated with it: its *rank* and its *dimension*. Its *rank* determines the value of k for which $b \in U_k$, and the dimension of b determines its dimension within this *chain-complex*.

The reader might wonder why we don't simply define $U_n = \text{Hom}_{\mathbb{Z}}(C, C^n)$ for a suitable chain-complex C and be done with it. It turns out that the added generality of our definition is absolutely necessary in applications. In fact it turns out to *rarely* be true that $U_n = \text{Hom}_{\mathbb{Z}}(C, C^n)$ in an actual example. In general, we will be concerned with formal coalgebras that *map* to formal coalgebras of the form $\{\text{Hom}_{\mathbb{Z}}(C, C^n)\}$[1] via maps that preserve composition-operations.

2.1.2. Note that $\{a \circ_k b\}$ denotes a *set* of composites of the two elements a, b, namely $\{a \circ_1 b, \ldots, a \circ_m b\}$. Here the rank of $b = m$ so the rightmost entry in this sequence is $a \circ_{\text{rank } b} b$

2.1.3. Multiple compositions are assumed to be *right-associative* unless otherwise stated — i.e. $a \circ_i b \circ_j c = a \circ_i (b \circ_j c)$.

2.1.4. A formal coalgebra will be called *unitary* if it contains an identity element with respect to the composition-operations $\{\circ_i\}$. This will clearly have to be an element of rank 1 and dimension 0.

---

[1] See the definition of the power construct — 2.4 on page 12.

DEFINITION 2.2. Let A and B be formal coalgebras. A morphism $f: A \to B$ is a morphism of the underlying chain-complexes, that preserves the composition operations.

Now we give a few examples of formal coalgebras:

DEFINITION 2.3. The *trivial* formal coalgebra, denoted I, is defined as follows:

1. its component of every rank n is the chain complex equal to $\mathbb{Z}$, concentrated in dimension 0;
2. there is one $\mathbb{Z}$-basis element $\{b_i\}$ of each rank — this is a generator of the zero-dimensional chain-module of the component of I of rank rank b_i.
3. the $\{b_i\}$ satisfy the composition-law: $b_i \circ_\alpha b_j = b_{i+j-1}$, for all values of α, which can run from 1 to j. The differential of this formal coalgebra is *identically zero*.

REMARK. 2.3.1. This is clearly a unitary formal coalgebra — the identity element is b_1.

DEFINITION 2.4. Let C be a DGA-module with augmentation $\epsilon: C \to \mathbb{Z}$, and with the property that $C_0 = \mathbb{Z}$. Then the *power-construct* of C, denoted $\mathbf{P}(C)$ is defined to be the formal coalgebra with:

1. component of rank $i = \operatorname{Hom}_{\mathbb{Z}}(C, C^i)$, with the differential induced by that of C and C^i. The *dimension* of an element of $\operatorname{Hom}_{\mathbb{Z}}(C, C^i)$ (for some i) is defined to be its *degree* as a map of graded $\mathbb{Z}$-modules (see 1.1 on page 9).
2. The $\mathbb{Z}$-summand is generated by one element, e, of rank 0.

Let $s_1 \in \operatorname{Hom}_{\mathbb{Z}}(C, C^i)$ and $s_2 \in \operatorname{Hom}_{\mathbb{Z}}(C, C^j)$ be elements of rank i and j, respectively, where $i, j \geq 1$. Then the composition $s_1 \circ_k s_2$, where $1 \leq k \leq j$, is defined by: $s_1 \circ_k s_2 = \underbrace{1 \otimes \cdots \otimes s_1 \otimes \cdots \otimes 1}_{k^{\text{th}} \text{ position}} \circ s_2: C \to C^{i+j-1}$. The composition $e \circ_k s_2$ is defined in a similar way, by identifying e with the augmentation map of C — it follows that $e \circ_k s_2 \in \operatorname{Hom}_{\mathbb{Z}}(C, C^{j-1})$, as one might expect.

For all i, $0 \leq i < \infty$, the canonical subcomplex $\operatorname{Hom}_{\mathbb{Z}}(C, C^i)$ of elements of rank i, is equipped with a natural S_i-action[2] — it is defined by permutation of the factors of the target, C^i.

Remarks.   2.4.1. This is nothing but the formal coalgebra described in the beginning of this section. The power construct is clearly *unitary* — its identity element is the identity map id $\in \operatorname{Hom}_{\mathbb{Z}}(C, C)$.

2.4.2. The motivation for formal coalgebras is that they model something like iterated coproducts that occur in the power-construct. We will use formal coalgebras as an convenient algebraic framework for defining other constructs that have topological applications.

2.4.3. The reader who is familiar with the notion of operads (see [15]) will notice that the power construct is an algebraic analogue to the dual of the *endomorphism operad* of a space.

---

[2] Here S_i represents the permutation group on i objects.

DEFINITION 2.5. Let C_1 and C_2 be formal coalgebras. Then $C_1 \otimes C_2$ is defined to have:

1. component of rank $i = (C_1)_i \otimes (C_2)_i$, where $(C_1)_i$ and $(C_2)_i$ are, respectively, the components of rank i of of C_1 and C_2. The tensor-product here is formed in the usual way for chain-complexes. If $a \in (C_1)_i$ and $b \in (C_2)_i$ are both of rank i, this implies that the *dimension* of $a \otimes b \in (C_1)_i \otimes (C_2)_i$ is $\dim(a) + \dim(b)$, and the *rank* of $a \otimes b = \text{rank}(a) = i$;
2. composition operations defined via

$$(a \otimes b) \circ_i (c \otimes d) = (-1)^{\dim(b)\dim(c)}(a \circ_i c \otimes b \circ_i d)$$

for $a, c \in C_1, b, d \in C_2$.

Here is an example that shows some of the expressive powers of formal coalgebras:

PROPOSITION 2.6. *Let C be a DGA-module. Co-associative coalgebra structures on C can be identified with morphisms $f \colon I \to \mathbf{P}(C)$, the the trivial formal coalgebra to the power construct of C.*

PROOF. Suppose we are given such a morphism $f \colon I \to \mathbf{P}(C)$. Define a coproduct $\Delta \colon C \to C^2$, by $\Delta = f(b_2)$ (in the notation of 2.3). The definition of the power construct, the fact that morphisms of formal coalgebras preserve all structures (i.e. rank, dimension, and composition operations) and the definition of the trivial formal coalgebra, imply that:

1. $\Delta \colon C \to C^2$ is a chain-map (since the differential of I is trivial);
2. $(1 \otimes \epsilon) \circ \Delta = (\epsilon \otimes 1) \circ \Delta = f(b_1) = id \colon C \to C$;
3. $(\Delta \otimes 1) \circ \Delta = f(b_2 \circ_1 b_2) = b_3 = f(b_2 \circ_2 b_2) = (1 \otimes \Delta) \circ \Delta$;

Consequently (C, Δ) is a co-associative coalgebra. The converse follows by a similar argument. ☐

We now define a very important formal coalgebra — the *symmetric construct*. It models the formal behavior of $\{\text{Hom}_{\mathbb{Z}}(C, C^n)\}$ in which each C^n is equipped with an action of the symmetric group, S_n that permutes the factors of C.

The symmetric construct will be denoted $\mathfrak{S}$. Its components are $\{\text{R}(S_n)\}_{n \in \mathbb{Z}+}$, where:

1. S_n denotes the symmetric group on n objects;
2. $\text{R}(S_n)$ denotes the bar-resolution of $\mathbb{Z}$ over $\mathbb{Z}S_n$;

Here we follow the convention that $\text{R}(S_0) = \text{R}(S_1) = \mathbb{Z}$, concentrated in dimension 0. *Pure* elements of $\mathfrak{S}$ are canonical basis elements of $\text{R}(S_n)$ for all values of n, or the generator 1 of the $\mathbb{Z}$-summand (of ranks 0 and 1). (By canonical basis elements, we mean elements of the form $[g_1| \dots |g_k] \in \text{R}(S_n)$).

Remarks. 2.6.1. This will turn out to be a *unitary* formal-coalgebra. Its identity element is the generator $[\ ] \in \text{R}(S_0)_0$. Note that a pure element $[g_1| \dots |g_k]$ is not determined in $\mathfrak{S}$ by the symmetric-group elements $g_1, \dots, g_k$ — one must also specify which summand, $\text{R}(S_n)$, the element resides in. We therefore distinguish between S_n and their isomorphic images in S_N, where $N > n$.

In order to describe the composition operations of $\mathfrak{S}$ we must define a few more terms:

DEFINITION 2.7. Let $\alpha_i, i = 1, \ldots, n$ be integers ≥ 0. Define a set-mapping $T_{\alpha_1, \ldots, \alpha_n} \colon S_n \to S_{|\alpha|}$, where $|\alpha| = \sum_{i=1}^n \alpha_i$ for a permutation $\sigma \in S_n$ as follows:

1. for i between 1 and n let L_i denote the sequence of integers: $\{A_i, A_i + 1, \ldots, A_i + \alpha_i - 1\}$ of length α_i, where $A_i = \sum_{j=1}^{i-1} \alpha_j$ — for instance the concatenation of all of the L_i is the sequence of integers from 1 to $|\alpha|$;
2. $T_{\alpha_1, \ldots, \alpha_n}(\sigma)$ is the permutation on the integers $1, \ldots, |\alpha|$ that permutes the blocks $\{L_i\}$ via σ. In other words, if σ is the permutation $\begin{pmatrix} 1 & \cdots & n \\ \sigma(1) & \cdots & \sigma(n) \end{pmatrix}$, then $T_{\alpha_1, \ldots, \alpha_n}(\sigma)$ is the permutation defined by writing $\begin{pmatrix} L_1 & \cdots & L_n \\ L_{\sigma(1)} & \cdots & L_{\sigma(n)} \end{pmatrix}$ and regarding the upper and lower rows as sequences length $|\alpha|$.

Remarks. 2.7.1. Do not confuse the T-maps defined here with the transposition map for tensor products of chain-complexes (see remark 1.2.4 on page 9). We will use the special notation T_i to represent $T_{1, \ldots, 2, \ldots, 1}$, where the 2 occurs in the ith position. The two notations don't conflict since the old notation is never used in the case when $n = 1$.

2.7.2. Here is an example of the computation of $T_{2,1,3}((1,3,2)) = T_{2,1,3}\begin{pmatrix} 1 & 2 & 3 \\ 3 & 1 & 2 \end{pmatrix}$:

$L_1 = \{1,2\}$, $L_2 = \{3\}$, $L_3 = \{4,5,6\}$. The permutation maps the ordered set $\{1,2,3\}$ to $\{3,1,2\}$, so we carry out the corresponding mapping of the sequences $\{L_1, L_2, L_3\}$ to get $\begin{pmatrix} L_1 & L_2 & L_3 \\ L_3 & L_1 & L_2 \end{pmatrix} = \begin{pmatrix} \{1,2\} & \{3\} & \{4,5,6\} \\ \{4,5,6\} & \{1,2\} & \{3\} \end{pmatrix} = \begin{pmatrix} 1 & 2 & 3 & 4 & 5 & 6 \\ 4 & 5 & 6 & 1 & 2 & 3 \end{pmatrix}$ (or $(1,4)(2,5)(3,6)$, in cycle notation).

It is possible to give a formula for computing $T_\alpha \sigma$ (where $\alpha = (\alpha_1, \ldots, \alpha_n)$):

DEFINITION 2.8. If i is an integer between 1 and $|\alpha| = \sum_{j=1}^n \alpha_j$, define

1. $z(i, \alpha, \sigma) = 1 + $ largest value of j such that $\sum_{k=1}^j \alpha_{\sigma(k)} < i$;
2. $d(i, a, \sigma) = i - \sum_{k=1}^{z(i,\alpha,\beta)} \alpha_k$;

Then

$$T_\alpha \sigma(i) = \sum_{k=1}^{\sigma(z(i,\alpha,\beta))} \alpha_k + d(i, a, \sigma)$$

PROOF. $z(i, \alpha, \sigma)$ is equal to the ordinal number of the list L_j containing the ith element of the concatenation mentioned in 2.7, above. $d(i, \alpha, \sigma)$ is equal to the distance of the ith element of the concatenation from the end of that list — it is usually a negative number. The formula for $T_\alpha \sigma(i)$ then computes the value of the list-element (in L_j) that occurs in this position. It essentially computes the last number occurring in this concatenation of lists and counts back from this to position i. $\square$

PROPOSITION 2.9. *Let σ_1 and σ_2 be two permutations of S_n and let $\{\alpha_1, \ldots, \alpha_n\}$ be any set of indices ≥ 0. Then*

$$\mathrm{T}_{\alpha_1, \ldots, \alpha_n}(\sigma_1 \cdot \sigma_2) = \mathrm{T}_{\alpha_1, \ldots, \alpha_n}(\sigma_1) \cdot \mathrm{T}_{\sigma_1^{-1}\{\alpha_1, \ldots, \alpha_n\}}(\sigma_2)$$

here the notation $\sigma_1^{-1}\{\alpha_1, \ldots, \alpha_n\}$ means "permute the list of n elements $\{\alpha_1, \ldots, \alpha_n\}$ via σ_1^{-1}".

PROOF. We use 2.8 on page 14. Note that $\sigma_1^{-1}\{\alpha_1, \ldots, \alpha_n\} = \{\alpha_{\sigma_1(1)}, \ldots, \alpha_{\sigma_1(n)}\}$. Now $z(i, \alpha, \sigma_2) = 1 + $ largest j such that $\sum_{k=1}^{j} \alpha_{\sigma_1\sigma_2(k)} < i$ — the subscript of the α is $\sigma_1\sigma_2(k)$ because the σ_2 is applied to the ordinal position of the α, and in $\{\alpha_{\sigma_1(1)}, \ldots, \alpha_{\sigma_1(n)}\}$ this is the argument of σ_1. It follows that $z(i, \sigma_1^{-1}\alpha, \sigma_2) = z(i, \alpha, \sigma_1\sigma_2)$. We get the following formula for $\mathrm{T}_{\sigma_1^{-1}\{\alpha_1, \ldots, \alpha_n\}}(\sigma_2)(i)$:

$$\sum_{k=1}^{\sigma_2(z(i,\alpha,\sigma_1\sigma_2))} \alpha_{\sigma_1(k)} + i - \sum_{k=1}^{z(i,\alpha,\sigma_1\sigma_2)} \alpha_{\sigma_1\sigma_2(k)}$$

It immediately follows (from the definition) that $z(\mathrm{T}_{\sigma_1^{-1}\{\alpha_1, \ldots, \alpha_n\}}(\sigma_2)(i), \alpha, \sigma_1) = \sigma_1(z(i, \alpha, \sigma_1\sigma_2))$. We finally conclude that

$$\mathrm{T}_{\alpha_1, \ldots, \alpha_n}(\sigma_1)\left(\mathrm{T}_{\sigma_1^{-1}\{\alpha_1, \ldots, \alpha_n\}}(\sigma_2)(i)\right) = \sum_{k=1}^{\sigma_1\sigma_2(z(i,\alpha,\sigma_1\sigma_2))} \alpha_k + i - \sum_{k=1}^{z(i,\alpha,\sigma_1\sigma_2)} \alpha_{\sigma_1\sigma_2(k)}$$

But this is precisely the formula for $\mathrm{T}_{\alpha_1, \ldots, \alpha_n}(\sigma_1 \cdot \sigma_2)(i)$. $\square$

This property of the T-maps makes it possible to extend them to chain maps of the corresponding bar-resolutions. We first recall how the differential is defined on the $\{\mathrm{R}(S_n)\}$.

The bar resolution may be regarded as a simplicial complex with face operations given by:

$$(2.1) \qquad \mathrm{F}_i[a_1|\ldots|a_m] = \begin{cases} a_1[a_2|\ldots|a_m] & \text{if } i = 0 \\ [a_1|\ldots|a_i \cdot a_{i+1}|\ldots|a_m] & \text{if } 0 < i < m \\ [a_1|\ldots|a_{m-1}] & \text{if } i = m \end{cases}$$

and, as usual, the differential is defined as an alternating sum of these face-operations.

It also has a well-known co-associative coalgebra structure given by $\Delta_R(a) = \sum_{i=0}^{n} \tilde{\mathrm{F}}^i(a) \otimes \mathrm{F}_0^{n-i}(a)$, where $\tilde{\mathrm{F}}$ is the last face operator. With the face operations given above this amounts to $\Delta_R: [a_1|\ldots|a_n] = \sum_{i=0}^{n}[a_1|\ldots|a_i] \otimes a_1 \cdots a_i[a_{i+1}|\ldots|a_n]$.

DEFINITION 2.10. Inductively define maps

$$\mathfrak{T}_{\alpha_1, \ldots, \alpha_n}: \mathrm{R}(S_n) \to \mathrm{R}(S_{|\alpha|})$$

by

$$\mathfrak{T}_{\alpha_1, \ldots, \alpha_n}([\sigma|a]) = \mathrm{T}_{\alpha_1, \ldots, \alpha_n}([\sigma])|\mathfrak{T}_{\sigma^{-1}\{\alpha_1, \ldots, \alpha_n\}}(a)]$$

and

$$\mathfrak{T}_{\alpha_1, \ldots, \alpha_n}(\sigma \cdot a) = \mathrm{T}_{\alpha_1, \ldots, \alpha_n}(\sigma) \cdot \mathfrak{T}_{\sigma^{-1}\{\alpha_1, \ldots, \alpha_n\}}(a)$$

where $a \in \mathrm{R}(S_n)$ and $s \in S_n$.

Proposition 2.9 on page 15 and a simple computation shows that the $\{\mathfrak{T}_{\alpha_1,\ldots,\alpha_n}\}$ preserve the face-operations given in formula 2.1. This immediately implies:

PROPOSITION 2.11. *The maps* $\mathfrak{T}_{\alpha_1,\ldots,\alpha_n}\colon \mathrm{R}(S_n) \to \mathrm{R}(S_{|\alpha|})$ *are chain-maps and homomorphisms of coalgebras.*

REMARK. 2.11.1. These chain-maps are generally only $\mathbb{Z}$-linear. Since they preserve face operations, they define geometric maps

$$\widetilde{K(S_n,1)} \xrightarrow{[\mathfrak{T}_{\alpha_1,\ldots,\alpha_n}]} \widetilde{K(S_{|\alpha|},1)}$$

of universal covering spaces of Eilenberg-MacLane spaces.

We will need one more algebraic construct in order to define the composition operations of $\mathfrak{S}$ — the *twisted shuffle product:*

DEFINITION 2.12. The twisted shuffle product, $\circledast\colon \mathrm{R}(S_n) \otimes \mathrm{R}(S_n) \to \mathrm{R}(S_n)$, is defined inductively by the following rules:

1. $a \cdot [a_1|\ldots|a_n] \circledast b \cdot [b_1|\ldots|b_m] = ab \cdot ([b^{-1}a_1b|\ldots|b^{-1}a_nb] \circledast [b_1|\ldots|b_m])$;
2. $1 \cdot [a_1|\ldots|a_n] \circledast 1 \cdot [b_1|\ldots|b_m] = \varphi((\partial[a_1|\ldots|a_n]) \circledast [b_1|\ldots|b_m]) + (-1)^n \varphi([a_1|\ldots|a_n] \circledast \partial[b_1|\ldots|b_m])$, where the a_i, a and the b_j, b are elements of S_n.

PROPOSITION 2.13. *The map* $\circledast\colon \mathrm{R}(S_n) \otimes \mathrm{R}(S_n) \to \mathrm{R}(S_n)$ *is a chain-map.*

PROOF. It is only necessary to note that, if we define $\circledast$ inductively, starting with elements of the form $1 \cdot [a_1|\ldots|a_n]$, and $1 \cdot [b_1|\ldots|b_m]$ in a given dimension and extend it to elements of the form $a \cdot [a_1|\ldots|a_n]$ using rule 1, the extension will be a chain-map because conjugation defines a chain map. The proof that we get a chain map in higher dimensions follows from the fact that φ is a contracting homotopy for $\mathrm{R}(S_n)$. $\square$

PROPOSITION 2.14. *The definition of the twisted shuffle product that appears above can be replaced by*

Let $[a_1|\ldots|a_n]$, $[b_1|\ldots|b_m] \in \mathrm{R}(S_k)$ *be canonical basis elements. The twisted shuffle product,* $[a_1|\ldots|a_n] \circledast [b_1|\ldots|b_m]$, *is an element of* $\mathrm{R}(S_k)$ *equal to*

$$(2.2) \quad (-1)^n [b_1([b_1^{-1}a_1b_1|\ldots|b_1^{-1}a_nb_1] \circledast [b_2|\ldots|b_m])] \\ + [a_1|([a_2|\ldots|a_n] \circledast [b_1|\ldots|b_m])$$

On $\mathbb{Z}$*-basis elements of* $\mathrm{R}(S_k)$ *of the form* $a \cdot [a_1|\ldots|a_n]$ *and* $b \cdot [b_1|\ldots|b_m]$ *(where* $a, b \in S_k$*), it is defined by* $ab \cdot ([b^{-1}a_1b|\ldots|b^{-1}a_nb] \circledast [b_1|\ldots|b_m])$.

REMARK. 2.14.1. The fact that the twisted shuffle product is associative and preserves the coproduct of the $\{\mathrm{R}(S_k)\}$ implies that each $\mathrm{R}(S_k)$ is a DGA-Hopf algebra.

PROOF. This follows immediately from the facts that:

1. $\varphi(a \cdot [a_1|\ldots|a_n]) = 0$ if $a = 1$;
2. the image of φ is an element of the form $a \cdot [a_1|\ldots|a_n]$ with $a = 1$, so that $1 \cdot [a_1|\ldots|a_n] \circledast 1 \cdot [b_1|\ldots|b_m]$ is a linear combination of elements of the form $c \cdot [c_1|\ldots|c_{n+m}]$ with $c = 1$.

It follows that the only terms of $\varphi((\partial 1 \cdot [a_1|\ldots|a_n]) \circledast 1 \cdot [b_1|\ldots|b_m]) + (-1)^n \varphi(1 \cdot [a_1|\ldots|a_n] \circledast \partial 1 \cdot [b_1|\ldots|b_m])$, the will contribute to the result will be those derived from the "tail terms" of the differential, ∂. $\square$

PROPOSITION 2.15. *The twisted shuffle product is associative.*

PROOF. This follows from 2.12 and induction on n and m. The statement is clear if any of the three factors is zero-dimensional. When this is not true the statement follows from the inductive hypothesis and the fact that we may take $\varphi \otimes 1$ as a contracting homotopy on $R(S_n) \otimes R(S_n)$ (at least above dimension 0). $\square$

PROPOSITION 2.16. *The twisted shuffle product preserves the coproduct.*

PROOF. This follows by an inductive argument like that used in 2.15. $\square$

We are now ready to define the *composition operations* of $\mathfrak{S}$:

PROPOSITION 2.17. *The graded differential module, $\mathfrak{S}$, as described above, equipped with the following composition operations, constitutes a formal coalgebra (in the sense of 2.1):*

$$\circ_i = \circledast(R(\mathcal{S}_{i-1}) \otimes \underbrace{\mathfrak{T}_{1,\ldots,m,\ldots,1}}_{i^{\text{th}}\ position}): R(S_m) \otimes R(S_n)) \rightarrow R(S_{m+n-1})$$

Here:

1. $\circledast$ *is the twisted shuffle product defined in 2.14;*
2. $\mathcal{S}_{i-1}$ *is the shift map $\mathcal{S}_{i-1}: S_m \rightarrow S_{m+i-1}$ — it shifts all indices of permutations up by $i-1$; and*
3. $R(\mathcal{S}_{i-1}): R(S_m) \rightarrow R(S_{m+i-1})$ *is the induced map of bar resolutions. Although $\mathcal{S}_{i-1}(R(S_m))$ lies in $R(S_{m+i-1})$, we will identify this with its image in $R(S_{m+n-1})$.*

Remarks. 2.17.1. This is proved in appendix A on page 99.

2.17.2. In many respects $\mathfrak{S}$ appears to be a kind of algebraic analogue to the *operad of little cubes*, $\mathcal{C}_1$, defined by May in [15]. This is a question that deserves further research.

PROPOSITION 2.18. *The coproducts $\{\Delta_n: R(S_n)) \rightarrow R(S_n) \otimes R(S_n)\}$ define a map of underlying chain-complexes $\Delta: \mathfrak{S} \rightarrow \mathfrak{S} \otimes \mathfrak{S}$ that is a morphism of formal coalgebras.*

REMARK. 2.18.1. This follows from:

1. the fact that the $\{\Delta_n\}$ preserve $\circledast$;
2. the fact that the $\{\Delta_n\}$ preserve $R(\mathcal{S}_{i-1})$-maps (clear);
3. the fact that the $\mathfrak{T}$-maps are homomorphisms of coalgebras (see 2.11 on page 16).

Note that 2.10 on page 15 implies that $\mathfrak{S}$ satisfies the identity:

PROPOSITION 2.19. *For all integers n, m, $0 \le n, m < \infty$, $a \in R(S_m)$, $b \in R(S_n)$, $g \in S_m, h \in S_n$, $(h \cdot b) \circ_i (g \cdot a) = \mathcal{S}_{j-1}(h)\underbrace{\mathfrak{T}_{1,\ldots,i,\ldots,1}}_{j^{\text{th}}\ position}(g) \cdot (b \circ_{g^{-1}(i)} a).$*

DEFINITION 2.20. A formal coalgebra $U = \{U_n\}$ will be said to be *symmetric* if, in addition to 2.1, it satisfies the conditions:

1. U_n is a left $\mathbb{Z}S_n$-module, for all n;
2. For all integers $n, m, 0 \leq n, m < \infty, a \in U_m, b \in U_n, g \in S_m, h \in S_n$, and all $1 \leq i \leq m, (h \cdot b) \circ_i (g \cdot a) = \mathcal{S}_{j-1}(h)\mathrm{T}_{\underbrace{1,\ldots,i,\ldots,1}_{j^{\text{th}} \text{ position}}}(g) \cdot \left(b \circ_{g^{-1}(i)} a\right).$

Remarks.   2.20.1. Condition 2 is very similar to the defining property of an operad in the category of chain-complexes.

2.20.2. Note that the power construct $\mathbf{P}(C)$ is symmetric if the nth component, $\mathrm{Hom}_{\mathbb{Z}}(C, C^n)$, is equipped with an action of S_n that permutes the factors of C_n, for all $n > 1$.

3. m-coalgebras

In this section we will actually define m-coalgebras. The study of these objects was originally motivated by a desire to compute a geometrically valid coproduct for the *cobar construction*. "Geometrically valid" in this context means a coproduct that is homotopic, as a map, to (say) that derived from the Alexander-Whitney diagonal map on the loop space. Although there is a very simple coproduct defined on the cobar construction (namely the shuffle coproduct), it is provably *not* geometric — see [19]. It quickly became clear that the coproduct on the cobar construction depended upon much more than that of the original chain complex — it depended nontrivially upon a "higher coproduct" like that used to define Steenrod operations. This process is complicated by the fact that some of the applications require that we work with m-coalgebras whose underlying chain complexes are infinitely generated or infinite-dimensional. The basic underlying idea is that we want to introduce a "higher diagonal" structure to the underlying chain complex. This is $\mathbb{Z}_2 = S_2$ equivariant chain-map:

$$\mathfrak{D}_2 \colon \mathrm{R}(S_2) \otimes C \to C \otimes C$$

like that defined by Davis in [6]. Here $\mathbb{Z}_2$ acts upon $C \otimes C$ by interchanging factors. It turns out that it is not enough to consider higher coproducts that map C to $C \otimes C$: we must also consider iterated coproducts or S_n-equivariant chain-maps

$$\mathfrak{D}_n \colon \mathrm{R}(S_n) \otimes C \to C^n$$

where n is some integer ≥ 1, $C^n = \underbrace{C \otimes \cdots \otimes C}_{n \text{ times}}$, and S_n acts on C^n by permuting factors. In general, the map $\mathfrak{D}_n$ contains data not dependent upon $\mathfrak{D}_2$ — although the co-associativity of the coproduct implies the existence of many relationships between different $\mathfrak{D}_i$. At this point, we could define the algebraic structures of interest to be *morphisms* of formal coalgebras $\mathfrak{S} \to \mathbf{P}(C)$, in analogy with 2.6. Here, we would also have to impose the additional requirement that the induced morphism of the nth component $\mathfrak{S}_n \to \mathbf{P}(C)_n$ is S_n equivariant, for all n. Such a definition satisfies most of our requirements. One problem with this definition is that the target, $\mathbf{P}(C)$ may

be *uncountably generated* in cases of interest, since C may fail to be finite-dimensional[3].
We resolve this particular problem by defining our algebraic structures to be maps

$$\mathfrak{D}'_n \colon C \to \operatorname{Hom}_{\mathbb{Z}S_n}(\mathrm{R}(S_n), C^n)$$

Note that this is essentially equivalent to the original formulation — we have merely
replaced a map by its *adjoint*. This doesn't completely resolve the difficulties, however.
In many cases of mathematical interest we will have to replace $\mathrm{R}(S_n)$ by a $\mathbb{Z}$-free chain
complex that is infinite dimensional. This will cause the targets of the maps $\mathfrak{D}'_n$ to be
uncountably generated. We would like a target that is finitely (or at least countably)
generated in each dimension if possible. This is not an unreasonable requirement, since
it turns out that the significant elements of the target of the maps $\mathfrak{D}'_n$ *generally* come
from finite-dimensional portions of the domains[4]. Our solution is to:

Adopt the second formulation of the algebraic structure — i.e.,
maps $\mathfrak{D}'_n \colon C \to \operatorname{Hom}_{\mathbb{Z}S_n}(\mathrm{R}(S_n), C^n)$.

With this in mind we define the chain-complexes to play the part of the $\mathrm{R}(S_n)$:

DEFINITION 3.1. An *f-resolution*, is a symmetric formal coalgebra $\mathfrak{R}$, that satisfies the
following conditions:

1. Each $\mathfrak{R}_n$ is a $\mathbb{Z}S_n$-resolution of $\mathbb{Z}$ and $\mathfrak{R}_1 = \mathfrak{R}_0 = \mathbb{Z}$ with trivial S_n-action.
2. Let e_2 be any element of $(\mathfrak{R}_2)_0$ whose image in $H_0(\mathfrak{R}_2) = \mathbb{Z}$ is 1. Let e_0 be
   $1 \in \mathbb{Z} = \mathfrak{R}_0$ and let e_1 be $1 \in \mathbb{Z} = \mathfrak{R}_1$. Then $e_0 \circ_i e_2 = e_1$, where $i = 1, 2$.
3. It has an augmentation map $\epsilon_n \colon (\mathfrak{R}_n)_0 \to \mathbb{Z}$.

DEFINITION 3.2. Given two f-resolutions $\mathcal{F}_1$ and $\mathcal{F}_2$, the tensor product $\mathcal{F}_1 \otimes \mathcal{F}_2$ is
defined to have an underlying formal coalgebra structure that is the tensor product of
the underlying formal coalgebras of $\mathcal{F}_1$ and $\mathcal{F}_2$.

We are now in a position to define *m-structures*

DEFINITION 3.3. Let C be a chain-complex with $H_0(C) = \mathbb{Z}$. Then:

1. An *m-structure* on C is defined to be a sequence of chain maps $\mathfrak{f}[C]_n \colon C \to
   \operatorname{Hom}_{\mathbb{Z}S_n}(\mathfrak{R}[C]_n, C^n)$, where $\mathfrak{R} = \{\mathfrak{R}[C]_n\}$ is some f-resolution, and n is an
   integer that satisfies $0 \leq n < \infty$. We assume that:
   a. the composite $e_1 \circ \mathfrak{f}_1 \colon C \to C^1$, is the identity map of C;
   b. and the composite $e_0 \circ \mathfrak{f}_0 \colon C \to C^0 = \mathbb{Z}$ coincides with the augmen-
      tation of C;
   c. For any $c \in C$, at most a finite number of the $\{\mathfrak{f}[C]_n(c)\}$ are nonzero.
      Here C^n is equipped with the S_n-action that permutes the factors.
   d. The adjoint will be denoted $\widetilde{\mathfrak{f}[C]}_n \colon \mathfrak{R}[C]_n \otimes C \to C^n$, and is defined
      by $\widetilde{\mathfrak{f}[C]}_n(r \otimes c) = (-1)^{\dim(r) \cdot \dim(c)} \mathfrak{f}[C]_n(c)(r)$, where $r \in \mathfrak{R}[C]_n$ and
      $c \in C$. With this definition in mind, we require $\widetilde{\mathfrak{f}[C]}_n(\mathfrak{R}[C]_n \otimes C(k)) \subseteq
      C(k)^n$, where $C(k)$ is the k-skeleton of C.
2. An m-structure will be called *weakly-coherent* if the *adjoint maps* fit into the
   commutative diagrams in figure 2.2.1 on page 20, for all $n, m \geq 1$ and $1 \leq
   i \leq m$.

---

[3] This happen when C is a cobar construction.
[4] I.e., the chain-complexes that replace the $\mathrm{R}(S_n)$.

JUSTIN R. SMITH

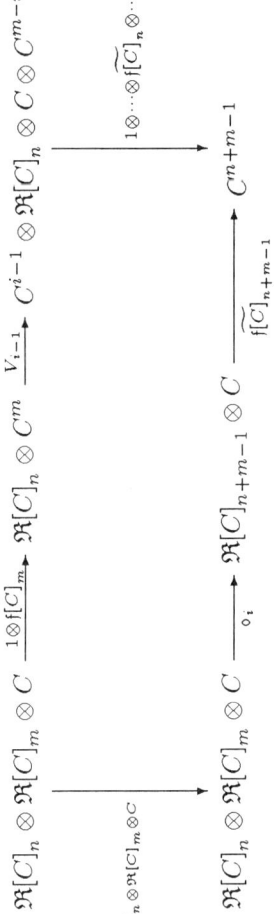

Figure 2.2.1. Adjoint maps in a weakly-coherent m-coalgebra

where $V\colon \mathfrak{R}[C]_n \otimes C^m \to C^{i-1} \otimes \mathfrak{R}[C]_n \otimes C \otimes C^{m-i}$ is the map that shuffles the factor $\mathfrak{R}[C]_n$ to the right of $i-1$ factors of C.

3. An m-structure $\{\mathfrak{f}[C]_n\colon C \to \operatorname{Hom}_{\mathbb{Z}S_n}(\mathfrak{R}[C]_n, C^n)\}$, will be called *strongly coherent* (or just *coherent*) if it is weakly coherent, and $\mathfrak{R}[C] = \mathfrak{S}$.

A chain-complex, C, equipped with an m-structure will be called an *m-coalgebra*. The maps $\mathfrak{f}[C]_n\colon C \to \operatorname{Hom}_{\mathbb{Z}S_n}(\mathfrak{R}[C]_n, C^n)$, where n is an integer such that $0 \le n < \infty$, will be called the *structure maps* of C.

Remarks. 3.3.1. If C is an incoherent m-coalgebra we may, without loss of generality, assume that $\mathfrak{R}[C] = \mathfrak{S}$, since the contracting homotopy, Φ, that is packaged with $\mathfrak{R}[C]$, allows us to construct a unique sequence of chain-map $\mathfrak{S}_n = \operatorname{R}(S_n) \to \mathfrak{R}[C]_n$, for n an integer such that $0 \le n < \infty$. We then compose the structure maps of the original m-coalgebra with the induced natural transformation $\operatorname{Hom}_{\mathbb{Z}S_n}(\mathfrak{R}[C], *) \to \operatorname{Hom}_{\mathbb{Z}S_n}(\mathfrak{S}, *)$, to get the structure maps of the modified m-coalgebra.

3.3.2. An m-coalgebra can be given the following interpretation: The adjoint isomorphism allows us to regard the structure maps as a family of S_n-equivariant chain-maps $\widetilde{\mathfrak{f}[C]}_n\colon \operatorname{R}(S_n) \otimes C \to C^n$. The map $\widetilde{\mathfrak{f}[C]}_2\colon \operatorname{R}(S_2) \otimes C \to C^2$, restricted to $[\,] \otimes C$, defines a kind of coproduct on C, called the *underlying coproduct* of the m-coalgebra. Define $\mathfrak{D}_a = \widetilde{\mathfrak{f}[C]}_i(a \otimes *)\colon C \to C^i$. These maps will be called the *higher-coproducts* associated with the m-coalgebra. The map $\mathfrak{D}_{[(1,2)]}\colon C \to C^2$ defines a chain-homotopy between $\Delta = \mathfrak{D}_{[\,]}$ and $T \circ \Delta$, where T is the transposition map defined in 1.2.5 on page 10.

3.3.3. The basic definitions in § 2 and § 3 can be stated in terms of *operads* in the category of graded differential modules. Operads were originally defined in terms of topological spaces by May in [15] and this concept was extended to DG-modules by Smirnov in [23]. Essentially:

1. the formal coalgebra $\mathfrak{S}$, constitutes an operad, and
2. a coherent m-coalgebra is a *coalgebra over* this operad, in the sense of § 3 of [23].

3.3.4. My original definition of an m-coalgebra regarded a coherent m-structure as a morphism of formal coalgebras $\mathfrak{S} \to \mathbf{P}(C)$, and a weakly coherent m-structure as a morphism $\mathfrak{R}[C] \to \mathbf{P}(C)$. Although this definition has the advantage of being much more elegant than the one given above it doesn't lend itself to effective computation unless C is finitely generated as a $\mathbb{Z}$-module — this means:

1. $C_i \ne 0$ for at most a finite number of values of i;
2. each of these nonzero C_i is, itself, finitely generated as a $\mathbb{Z}$-module.

3.3.5. The definition of weak coherence of an m-structure can be re-stated in terms of the maps $\{\mathfrak{f}[C]_n\}$ themselves, rather than their *adjoints* $\{\widetilde{\mathfrak{f}[C]}_n\}$. An m-structure is weakly coherent if and only if the diagram in figure 2.2.2 commutes for all integers n

such that $0 \leq n < \infty$. In this diagram, the map V_i' represents the composite

$$(3.1) \quad \mathrm{Hom}_{\mathbb{Z}S_n}(\mathfrak{R}[C]_n, C^{i-1} \otimes \mathrm{Hom}_{\mathbb{Z}S_m}(\mathfrak{R}[C]_m, C^m) \otimes C^{n-i})$$

$$\xrightarrow{i_1} \mathrm{Hom}_{\mathbb{Z}S_n}(\mathfrak{R}[C]_n, C^{i-1} \otimes \mathrm{Hom}_{\mathbb{Z}S_m}(\mathfrak{R}[C]_m, C^m) \otimes C^{n-i})$$

$$\xrightarrow{\mathrm{Hom}_{\mathbb{Z}S_n}(1, i_2)} \mathrm{Hom}_{\mathbb{Z}S_n}(\mathfrak{R}[C]_n, C^{i-1} \otimes \mathrm{Hom}_{\mathbb{Z}S_m}(\mathfrak{R}[C]_m, C^m) \otimes C^{n-i})$$

$$\rightarrow \mathrm{Hom}_{\mathbb{Z}}(\mathfrak{R}[C]_n \otimes \mathfrak{R}[C]_m, C^{n+m-1})$$

where i_1 and i_2 are inclusion mappings of the $\mathrm{Hom}_{\mathbb{Z}S_n}$-functors in the respective $\mathrm{Hom}_{\mathbb{Z}}$-groups. We are also including $\mathrm{Hom}_{\mathbb{Z}S_i}(*, *)$ in $\mathrm{Hom}_{\mathbb{Z}}(*, *)$, by simply forgetting that the elements are $\mathbb{Z}S_i$ linear.

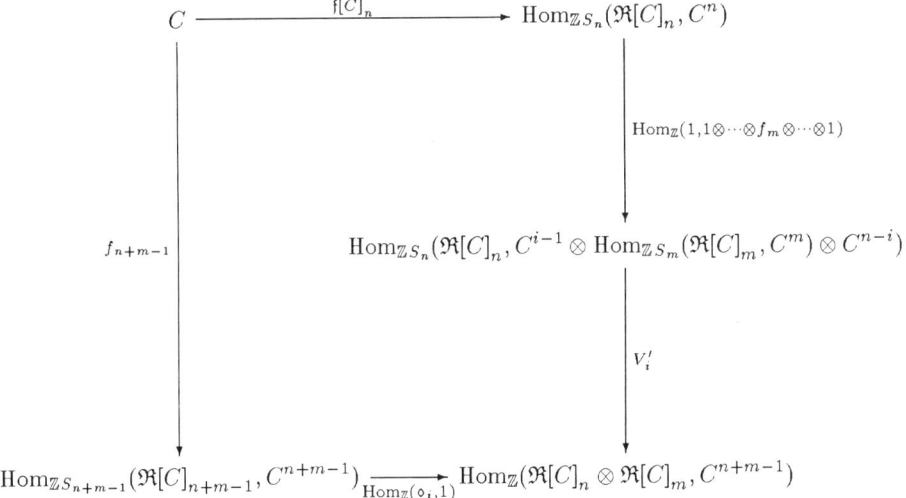

Figure 2.2.2. Structure maps of a weakly coherent m-coalgebra

This diagram means that the composition-operations in the coordinate coalgebra correspond to actual compositions of the adjoint maps.

Coherence of an m-structure implies that compositions of higher-coproducts are canonically equal to certain other higher coproducts. Thus, for instance, $\mathfrak{D}_{[(1,2)]} \otimes 1 \circ \mathfrak{D}_{[(1,2)]} = \mathfrak{D}_{[(1,2)] \circledast \mathfrak{T}_{2,1}[(1,2)]} = \mathfrak{D}_{[(1,3,2)] \circledast [(1,2)]} = \mathfrak{D}_{[(1,3,2)]|(1,2)] - [(1,2)|(1,2,3)]} = \mathfrak{D}_{[(1,3,2)]|(1,2)]} - \mathfrak{D}_{[(1,2)|(1,2,3)]}$. We can, consequently, translate any formula involving compositions of higher-coproducts into a formula involving elements of the $\{R(S_n)\}$.

The underlying coproduct of a coherent m-coalgebra defined by $\mathfrak{f}[C]_n \colon C \rightarrow \mathrm{Hom}_{\mathbb{Z}S_n}(\mathfrak{S}_n, C^n)$, where n is an integer such that $0 \leq n < \infty$ is co-associative — recall that $\mathfrak{S}_n = R(S_n)$. This follows from:

1. the fact that $\mathfrak{S}$ contains an embedded copy of the trivial formal coalgebra I;
2. $\mathfrak{f}(I)$ consists of all composites of the underlying coproduct of C;
3. Proposition 2.6 on page 13.

A similar argument implies that the underlying coproduct of a weakly coherent m-coalgebra is homotopy co-associative in a fairly strong sense. Given this interpretation of an m-structure in terms of coproducts and higher coproducts, we can define Steenrod operations:

DEFINITION 3.4. Let p be a prime and let C be an m-coalgebra with structure maps $\mathfrak{f}[C]_n \colon C \to \operatorname{Hom}_{\mathbb{Z}S_n}(\mathrm{R}(S_n), C^n)$ and associated higher coproduct $\hat{f}_n \colon \mathrm{R}(S_n) \otimes C \to C^n$, where n is an integer such that $0 \leq n < \infty$. If $\alpha \in H_k(C, \mathbb{Z}_p)$ is a cohomology class we can define $P^i(\alpha)$, the ith Steenrod p^{th} power operation, to be the class $(\widetilde{\mathfrak{f}[C]}_{e_{kp-i-k}})^* \underbrace{\alpha \times \cdots \times \alpha}_{p \text{ factors}} \in H^{k+i}(C, \mathbb{Z}_p)$, where:

1. $\underbrace{\alpha \times \cdots \times \alpha}_{p \text{ factors}} \colon (C)^p \to (\mathbb{Z}_p)^p = \mathbb{Z}_p$, is the external product;

2. $e_j \in \mathrm{R}(S_p)_j$ is defined inductively by $e_0 = [\ ]$, and

$$e_{i+1} = \begin{cases} \varphi(t \cdot e_i) & \text{if } i \text{ is even} \\ \sum_{m=1}^{p-1} \varphi(t^m \cdot e_i) & \text{if } i \text{ is odd} \end{cases}$$

where t is the cycle $(1, \ldots, p) \in S_p$ and φ is the contracting homotopy defined in 3.1.1 on page 19.

Remarks. 3.4.1. By abuse of notation, we have written α for a cycle representative of the class $\alpha \in H_k(C, \mathbb{Z}_p)$. It is not hard to see that the process of taking p-fold external products is linear if we are working mod p. The standard proof that Steenrod p^{th} powers are well defined shows that $P^i(\alpha)$ only depends on the class of α.

3.4.2. The elements $\{e_j\}$ simply define a mapping from a free resolution of $\mathbb{Z}$ over $\mathbb{Z}[\mathbb{Z}_p]$ into $\mathrm{R}(S_p)$.

3.4.3. Note that in the coherent case many higher diagonal structures determine each other — i.e. it is as if Steenrod squares determined many higher Steenrod p^{th} powers — at least if they are defined on the chain level. It doesn't appear that chain-level Steenrod squares determine all higher powers — the maps $\mathfrak{T}_{\alpha_1, \ldots, \alpha_n}$, with all $\alpha_i \geq 1$, for instance, are never surjective.

PROPOSITION 3.5. *Let* $\mathfrak{R}_1 = \{\mathfrak{R}_{1,n}\}$ *and* $\mathfrak{R}_2 = \{\mathfrak{R}_{2,n}\}$ *be* $\mathfrak{f}$-*resolutions, and let* C_1 *and* C_2 *be chain-complexes. Then there exists a natural transformation of functors* $\mathfrak{E}_n \colon \operatorname{Hom}_{\mathbb{Z}S_n}(\mathfrak{R}_{1,n}, C_1^n) \otimes \operatorname{Hom}_{\mathbb{Z}S_n}(\mathfrak{R}_{2,n}, C_2^n) \to \operatorname{Hom}_{\mathbb{Z}S_n}(\mathfrak{R}_{1,n} \otimes \mathfrak{R}_{2,n}, (C_1 \otimes C_2)^n)$, *for all* n.

REMARK. 3.5.1. If $u \in \operatorname{Hom}_{\mathbb{Z}S_n}(\mathfrak{R}_{1,n}, C_1^n)$, $v \in \operatorname{Hom}_{\mathbb{Z}S_n}(\mathfrak{R}_{2,n}, C_2^n)$, then $\mathfrak{E}_n$ sends $u \otimes v$ to $(c_1 \otimes c_2 \to V_n((u \otimes v)(c_1 \otimes c_2)))$, where $c_1 \in C_1, c_2 \in C_2$ and $V_n \colon C_1^n \otimes C_2^n \to (C_1 \otimes C_2)^n$ is the map that shuffles the factors of together.

We can now define tensor products of m-coalgebras:

DEFINITION 3.6. Let C_1 and C_2 be m-coalgebras with structure maps $\mathfrak{f}[C_i]_n \colon C_i \to \operatorname{Hom}_{\mathbb{Z}S_n}(\mathfrak{R}[C_i]_n, C_i^n)$, $i = 1, 2$ and $0 \leq n < \infty$. Their *tensor product* is an m-coalgebra whose underlying chain-complex is $C_1 \otimes C_2$ and whose structure map is the composite

$$C_1 \otimes C_2 \xrightarrow{\mathfrak{f}[C_1] \otimes \mathfrak{f}[C_2]} \operatorname{Hom}_{\mathbb{Z}S_n}(\mathfrak{R}[C_1]_n, C_1^n) \otimes \operatorname{Hom}_{\mathbb{Z}S_n}(\mathfrak{R}[C_2]_n, C_2^n)$$

$$\xrightarrow{\mathfrak{E}_n} \operatorname{Hom}_{\mathbb{Z}S_n}(\mathfrak{R}[C_1]_n \otimes \mathfrak{R}[C_2]_n, (C_1 \otimes C_2)^n)$$

Remarks.   3.6.1. Here $\mathfrak{R}[C_1] \otimes \mathfrak{R}[C_2]$ is the tensor product of f-resolutions, defined in 3.2, and $\mathfrak{C}_n \colon \mathrm{Hom}_{\mathbb{Z}S_n}(\mathfrak{R}[C_1]_n, C_1^n) \otimes \mathrm{Hom}_{\mathbb{Z}S_n}(\mathfrak{R}[C_2]_n, C_2^n) \to \mathrm{Hom}_{\mathbb{Z}S_n}(\mathfrak{R}[C_1]_n \otimes \mathfrak{R}[C_2]_n, (C_1 \otimes C_2)^n)$ is the map defined in 3.5.

3.6.2. It is clear that the tensor product of two weakly coherent m-coalgebras is also weakly coherent. Proposition 2.18 on page 17 implies that we can define tensor products of strongly coherent m-coalgebras in such a way that the corresponding statement is true:

DEFINITION 3.7. Let C_1 and C_2 be coherent m-coalgebras with sets of structure morphisms $\{\mathfrak{f}[C_i]_n \colon C_i \to \mathrm{Hom}_{\mathbb{Z}S_n}(\mathrm{R}(S_n), C_i^n)\}$, $i = 1, 2$, $0 \le n < \infty$. Their *tensor product* is a coherent m-coalgebra whose underlying chain-complex is $C_1 \otimes C_2$ and whose structure morphisms are the composites

$$C_1 \otimes C_2 \xrightarrow{\;\mathfrak{f}[C_1] \otimes \mathfrak{f}[C_2]\;} \mathrm{Hom}_{\mathbb{Z}S_n}(\mathrm{R}(S_n), C_1^n) \otimes \mathrm{Hom}_{\mathbb{Z}S_n}(\mathrm{R}(S_n), C_2^n)$$

$$\Big\downarrow \mathfrak{C}_n$$

$$\mathrm{Hom}_{\mathbb{Z}S_n}(\mathrm{R}(S_n), (C_1 \otimes C_2)^n) \xleftarrow{\;\mathrm{Hom}_{\mathbb{Z}S_n}(\Delta_R, 1)\;} \mathrm{Hom}_{\mathbb{Z}S_n}(\mathrm{R}(S_n) \otimes \mathrm{R}(S_n), (C_1 \otimes C_2)^n)$$

for all n.

Our treatment of m-coalgebras would be incomplete without a description of morphisms of such objects.

DEFINITION 3.8. Let C_1 and C_2 be m-coalgebras with sets of structure maps $\{\mathfrak{f}[C_i]_n \colon C_i \to \mathrm{Hom}_{\mathbb{Z}S_n}(\mathfrak{R}[C_i]_n, C_i^n)\}$, $i = 1, 2$, and all $0 \le n < \infty$. A *strict morphism* $\{g, h\} \colon C_1 \to C_2$ consists of:

1. a chain-map from $g \colon C_1 \to C_2$;
2. a morphism of f-resolutions, $h \colon \mathfrak{R}[C_2] \to \mathfrak{R}[C_1]$ such that the diagram

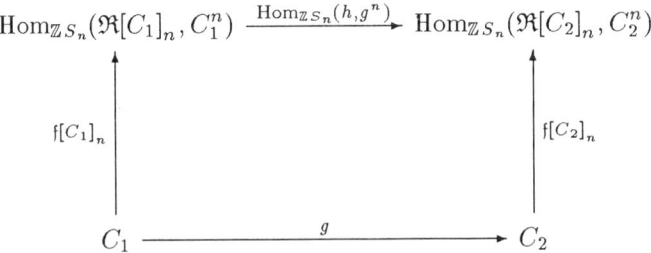

commutes for all n.

REMARK. 3.8.1. It is not hard to see that weakly-coherent m-coalgebras, equipped with strict morphisms, form a category, which will be denoted $\mathfrak{M}_0$.

A *contraction* of chain-complexes $(p, \iota, \varphi): C \rightarrow D$ is a pair of maps $p: C \rightarrow D$, $\iota: D \rightarrow C$ and a chain-homotopy $\varphi: C \rightarrow C$ such that $p \circ \iota = 1_D$, $\iota \circ p - 1_C = \partial \varphi$. The map p is called the *projection* of the contraction and ι is called its *injection* — see [10].

DEFINITION 3.9. Let C and D be weakly-coherent m-coalgebras. A contraction $(p, \iota, \varphi): C \rightarrow D$ with the injection, ι, a strict morphism of m-coalgebras, will be called an *elementary morphism* from $C \rightarrow D$.

Remarks.   3.9.1. Note that we have weakened the definition of morphism considerably — we regard formal inverses of strict morphisms that are injection of contractions to also be morphisms.

3.9.2. It is interesting to consider just what the definition of morphism means. Since projections of contractions are chain-maps, we can still regard a morphism as having an underlying map of chain-complexes. What we have relaxed in our definition of a morphism, is the requirement that all of the auxiliary structures be preserved. In a morphism of weakly-coherent m-coalgebras, these structures will generally only be preserved "up to a chain-homotopy", in some sense.

DEFINITION 3.10. The category of weakly-coherent m-coalgebras, denoted $\mathfrak{M}$, is defined to be the localization of $\mathfrak{M}_0$ by the set of strict morphisms whose associated chain-maps of underlying chain-complexes are injections of contractions of chain-complexes.

Remarks.   3.10.1. The objects of this category are weakly-coherent m-coalgebras as before, but the morphisms (in principal) are formal composites of strict morphisms and elementary morphisms as defined in 3.9. In actual fact, we may regard all morphisms as elementary since strict morphisms can also be regarded as elementary morphisms — just take the algebraic mapping cylinder of a strict morphism. A morphism whose underlying map of chain-complexes is a chain-homotopy equivalence will be called an *equivalence*.

3.10.2. See 4.5 on page 31 for some motivation for this definition. The definition is essentially set up so that the maps in the Eilenberg-Zilber theorem are morphisms.

DEFINITION 3.11. Let C and D be two weakly coherent m-coalgebras. A *morphism* is a contraction $(\varphi, f, g): P \rightarrow D$, with P a weakly coherent m-coalgebra, equipped with strict morphisms $\iota: C \rightarrow P$, and $g: D \rightarrow P$. The composite $f \circ \iota: C \rightarrow D$ is called the *underlying map* of the morphism. We will denote this morphism by $(\iota, \varphi, f, g): C \rightarrow P \rightarrow D$.

Morphisms preserve m-structures up to a chain-homotopy, as the following lemma shows:

LEMMA 3.12. *Let C_1 and C_2 be   m-coalgebras   with   structure   maps* $\mathfrak{f}[C_i]_n: C_i \rightarrow \mathrm{Hom}_{\mathbb{Z}S_n}(\mathfrak{R}[C_i]_n, C_i^n)$, $i = 1, 2$, *respectively. Let* $(\varphi, u, g): C_1 \rightarrow C_2$ *be a contraction with the property that the map* $(h, g): C_2 \rightarrow C_1$ *is a strict morphism of m-coalgebras (where* $h: \mathfrak{R}[C_1] \rightarrow \mathfrak{R}[C_2]$ *is the corresponding morphism of f-resolutions).*

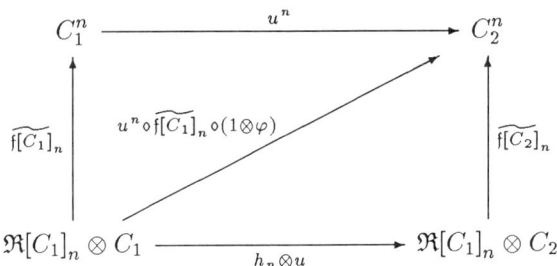

FIGURE 2.2.3. Injection morphism in a contraction

Then the diagram in figure 2.2.3 commutes up to a chain-homotopy (given in the diagonal map).

Consequently, the map u preserves m-structures up to a chain-homotopy — at least after the coordinate coalgebra of C_2 has been replaced by $\mathfrak{R}[C_1]$.

PROOF. $\partial(\mathrm{Hom}_{\mathbb{Z}S_n}(1, u^n) \circ \mathfrak{f}[C_1] \circ \varphi) = \mathrm{Hom}_{\mathbb{Z}S_n}(1, u^n) \circ \mathfrak{f}[C_1] \circ \partial\varphi$ (since $\mathrm{Hom}_{\mathbb{Z}S_n}(1, u^n) \circ \mathfrak{f}[C_1]$ is a chain-map) $= \mathrm{Hom}_{\mathbb{Z}S_n}(1, u^n) \circ \mathfrak{f}[C_1] \circ (g \circ u - 1)$ (by the property of a contraction). The fact that g is a strict morphism of m-coalgebras (see 3.8) implies that $\mathfrak{f}[C_1] \circ g = \mathrm{Hom}_{\mathbb{Z}S_n}(h, g_n) \circ \mathfrak{f}[C_2]$, which implies that $\partial(\mathrm{Hom}_{\mathbb{Z}S_n}(1, u^n) \circ \mathfrak{f}[C_1] \circ \varphi) = \mathrm{Hom}_{\mathbb{Z}S_n}(1, u^n) \circ \mathfrak{f}[C_1] \circ (g \circ u - 1) = \mathrm{Hom}_{\mathbb{Z}S_n}(1, u^n) \circ \mathrm{Hom}_{\mathbb{Z}S_n}(h, g_n) \circ \mathfrak{f}[C_2] \circ u - \mathrm{Hom}_{\mathbb{Z}S_n}(1, u^n) \circ \mathfrak{f}[C_1] = \mathrm{Hom}_{\mathbb{Z}S_n}(h, 1) \circ \mathfrak{f}[C_2] \circ u - \mathrm{Hom}_{\mathbb{Z}S_n}(1, u^n) \circ \mathfrak{f}[C_1]$, which is what we want to prove. $\square$

DEFINITION 3.13. Let $C = (C, \{\mathfrak{f}[C]_n : C \to \mathrm{Hom}_{\mathbb{Z}S_n}(\mathfrak{R}[C]_n, C^n)\})$ be a weakly-coherent m-coalgebra with structure morphisms $\mathfrak{f}[C]_n : C \to \mathrm{Hom}_{\mathbb{Z}S_n}(\mathfrak{R}[C]_n, C^n)$. Assume, in addition that, C, is equipped with the structure of a DGA-algebra as well, $\mu : C \otimes C \to C$. Then C will be called a *weakly-coherent m-Hopf* (respectively *strict weakly-coherent m-Hopf algebra*) algebra if:

1. for all $n > 0$, $\mathfrak{R}[C]_n$ has the structure of a DGA-coalgebra with coproduct $\Delta_{\mathfrak{R}[C]_n} : \mathfrak{R}[C]_n \to \mathfrak{R}[C]_n \otimes \mathfrak{R}[C]_n$;
2. The multiplication mapping $\mu : C \otimes C \to C$ is the underlying map of a morphism (respectively strict morphism) of weakly coherent m-coalgebras: $(\iota, \varphi, f, g) : P \to C$ (respectively strict morphism) of m-coalgebras, where the h-map (see statement 2 of 3.8) is $\Delta_{\mathfrak{R}[C]_n}$.
3. In addition the identity element (i.e. basepoint) of C has the property that

$$\mathfrak{f}[C]_n \circ \mathfrak{D}(P) \circ 1 \otimes \varphi \circ 1 \otimes \iota | \mathfrak{R}[C]_n \otimes 1 \otimes C = 0$$
$$= \mathfrak{f}[C]_n \circ \mathfrak{D}(P) \circ 1 \otimes \varphi \circ 1 \otimes \iota | \mathfrak{R}[C]_n \otimes C \otimes 1$$

where $\mathfrak{D}(P)$ is the (higher) coproduct mapping of P.

We will use the notation $(C, \mu[C], \{\mathfrak{f}[C]_n : C \to \mathrm{Hom}_{\mathbb{Z}S_n}(\mathfrak{R}[C]_n, C^n)\})$ for the m-Hopf algebra in this definition.

Remarks. 3.13.1. Note that $\partial(f_n \circ \mathfrak{D}(P) \circ 1 \otimes \varphi \circ 1 \otimes \iota) = \mathfrak{D}(A) \circ 1 \otimes \mu - \mu^n \circ \mathfrak{D}(A \otimes A)$, measures the extent to which μ preserves m-structures — see 3.12. The last condition simply states that this homotopy vanishes at the identity element.

3.13.2. Strict morphisms of m-Hopf algebras are defined to preserve all algebraic structures — including the coproducts $\mathfrak{D}_{\mathfrak{R}[C],n}: \mathfrak{R}[C]_n \to \mathfrak{R}[C]_n \otimes \mathfrak{R}[C]_n$. Morphisms are defined in analogy with morphisms for m-coalgebras except that:

1. all intermediate complexes must be weakly-coherent m-Hopf algebras and;
2. projections of intermediate contractions must be DGA-algebra homomorphisms (as before, the corresponding injections must be strict morphisms of m-Hopf algebras).

3.13.3. Condition 3 implies, roughly speaking, that the product-mapping is a "strict morphism at the identity element".

Since m-coalgebras can now have an algebra structure, we can define modules over m-Hopf algebras.

DEFINITION 3.14. Let $(A, \mu[A], \{f[A]_n : A \to \mathrm{Hom}_{\mathbb{Z}S_n}(\mathfrak{R}[A]_n, A^n)\})$ be a m-Hopf algebra, and let $(C, \{f[C]_n : C \to \mathrm{Hom}_{\mathbb{Z}S_n}(\mathfrak{R}[C]_n, C^n)\})$ be a weakly-coherent m-coalgebra. Then C will be called a (left) *strict m-module* over A if it is equipped with a strict morphism $A \otimes C \xrightarrow{\ a\ } C$ such that the corresponding map of coordinate-coalgebras (see 3.8 on page 24):

$$\mathfrak{R}[C] \xrightarrow{\ \mathfrak{R}[a]\ } \mathfrak{R}[A] \otimes \mathfrak{R}[C]$$

fits into the commutative diagram:

$$
\begin{array}{ccccc}
\mathfrak{R}[C] & \xrightarrow{\ \mathfrak{R}[a]\ } & \mathfrak{R}[A] \otimes \mathfrak{R}[C] & \xrightarrow{\ 1 \otimes a\ } & \mathfrak{R}[A] \otimes \mathfrak{R}[A] \otimes \mathfrak{R}[C] \\
\Big\| & & & & \Big\| \\
\mathfrak{R}[C] & \xrightarrow[\ \mathfrak{R}[a]\ ]{} & \mathfrak{R}[A] \otimes \mathfrak{R}[C] & \xrightarrow[\ 1 \otimes \Delta_{\mathfrak{R}[A]}\ ]{} & \mathfrak{R}[A] \otimes \mathfrak{R}[A] \otimes \mathfrak{R}[C]
\end{array}
$$

Here $\Delta_{\mathfrak{R}[A]}$ is the coproduct of $\mathfrak{R}[A]$. The fact that $\mathfrak{R}[A]$ is a coalgebra follows from the definition of an m-Hopf algebra (see 3.13 above).

We can also define *non*-strict m-modules:

DEFINITION 3.15. Let $(A, \mu[A], \{f[A]_n : A \to \mathrm{Hom}_{\mathbb{Z}S_n}(\mathfrak{R}[A]_n, A^n)\})$ be a m-Hopf algebra, and let F be a right DGA-module over A, with action-mapping $a: F \otimes A \to F$. Then F is called a (right)*m-module* (respectively strict right m-module) over A if:

1. F is a weakly-coherent m-coalgebra $(F, \{f[F]_n : F \to \mathrm{Hom}_{\mathbb{Z}S_n}(\mathfrak{R}[F]_n, F^n)\})$;

2. the map $a\colon A \otimes F \to F$, defining the action of A on F is a morphism $(\iota, \varphi, f, g)\colon P \to F$ of m-coalgebras such that corresponding adjoint morphism of f-resolutions (or g-map — see 3.11 on page 25) $\mathfrak{C}_n\colon \mathfrak{R}[F]_n \to \mathfrak{R}[F]_n \otimes \mathfrak{R}[A]_n$ defines $\mathfrak{R}[F]_n$ to be a right $\mathfrak{R}[A]_n$-comodule, for all n.

3. If the map $a\colon A \otimes F \to F$ is a non strict morphism, $(\iota, \varphi, f, g)\colon A \otimes F \to P \to F$, then $f_n \circ \mathfrak{D}(P) \circ 1 \otimes \varphi \circ 1 \otimes \iota | \mathfrak{R}[A]_n \otimes 1 \otimes F = 0$, where $\mathfrak{D}(P)$ is the (higher) coproduct mapping of P.

REMARK. 3.15.1. The definition of a right comodule implies that the following diagram commutes:

$$
\begin{array}{ccccc}
\mathfrak{R}[F] & \xrightarrow{\ \mathfrak{R}[a]\ } & \mathfrak{R}[A] \otimes \mathfrak{R}[F] & \xrightarrow{\ 1 \otimes a\ } & \mathfrak{R}[A] \otimes \mathfrak{R}[A] \otimes \mathfrak{R}[F] \\
\Big\| & & & & \Big\| \\
\mathfrak{R}[F] & \xrightarrow[\ \mathfrak{R}[a]\ ]{} & \mathfrak{R}[A] \otimes \mathfrak{R}[F] & \xrightarrow[\ 1 \otimes \Delta_{\mathfrak{R}[A]}\ ]{} & \mathfrak{R}[A] \otimes \mathfrak{R}[A] \otimes \mathfrak{R}[F]
\end{array}
$$

These concepts will be particularly important in the study of the m-structures of fibrations — see § 2 on page 86.

4. m-coalgebras and topological spaces

In this section we relate the algebraic constructs of the previous three sections to topology. We define a functor on CW-complexes taking its values in coherent m-coalgebras.

LEMMA 4.1. *Let C be a DG-module that is acyclic in positive dimensions, with split augmentation map $\epsilon\colon C_0 \to \mathbb{Z} \subset C_0$, and a self-annihilating contracting chain-homotopy $\varphi\colon C \to C$. Let $\Phi_n = 1 \otimes \cdots \otimes \varphi + 1 \otimes \cdots \otimes \varphi \otimes \epsilon + \cdots + \varphi \otimes \cdots \otimes \epsilon$ (n factors in each term). For $k > 1$ suppose $C(k)$ denotes the k-skeleton of C and $f_n\colon C(k) \to \operatorname{Hom}_{\mathbb{Z}S_n}(\mathrm{R}(S_n), C(k)^n)$ is defined for all $n \geq 1$ and:*

For all canonical basis elements $\sigma \in \mathrm{R}(S_n)$, $c \in C(k)$ with $c \in \operatorname{im}\varphi$,
$f_n(c)(\sigma) \subset \operatorname{im}\Phi_n$.

Then the $\{f_n\}$ can be extended to a map $f_n\colon C(k+1) \to \operatorname{Hom}_{\mathbb{Z}S_n}(\mathrm{R}(S_n), C(k+1)^n)$ in the image of φ[5] by an inductive application of the formula: $f_n(c)(\sigma) = \Phi_n \circ f_n(\partial c)(\sigma) + (-1)^{\dim(c)}\Phi_n \circ f_n(c)(\partial\sigma)$, where $c \in C_{k+1}$. This extension of the m-structure on $C(k)$ to $C(k+1)$ will be coherent.

Remarks.   4.1.1. Canonical basis elements of $\mathrm{R}(S_n)$ are, of course, elements of the form $[g_1| \ldots |g_i]$, where $g_1, \ldots, g_i \in S_n$. Defining the m-structure on canonical basis elements defines it on all of $\mathrm{R}(S_n)$ since we assume that f_n is $\mathbb{Z}S_n$-linear and S_n acts on C^n via permutation of the factors.

4.1.2. It turns out that the statement of the result is also true if we define $\Phi_n = \varphi \otimes \cdots \otimes 1 + \epsilon \otimes \varphi \otimes \cdots \otimes 1 + \cdots + \epsilon \otimes \cdots \otimes \varphi$.

---

[5]In other words, $f_n|\varphi(C_k)$ will be well-defined.

4.1.3. This lemma is derived from the Cartan Theory of Constructions — see [5].

PROOF. Let n and m be an arbitrary integers $0 \leq n, m < \infty$. We need to verify the coherency condition 3.10 on elements of C_{k+1}, paired with canonical basis elements of $R(S_n)$. This coherency condition is:

Let $c \in C_{k+1}$ be in the image of φ. $f_m(C) \circ_i (f_n(C)(c)(\sigma_1))(\sigma_2) = f_{n+m-1}(C)(c)(\circledast(R(S_{i-1}) \otimes \underbrace{\mathfrak{T}_{1,\ldots,m,\ldots,1}}_{i^{\text{th}} \text{ position}}))(\sigma_1 \otimes \sigma_2)$, where the $\sigma_1 \in$ $R(S_m)$, and $\sigma_2 \in R(S_n)$ are canonical basis elements. Here, the notation $f_m(C) \circ_i \ldots$ means $\underbrace{1 \otimes \cdots \otimes f_m(C) \otimes \cdots \otimes 1}_{i^{\text{th}} \text{ position}}$.

Claim 1: Both sides of this equation have their image in the image of Φ_{n+m-1}. This claim follows from:

1. condition 2 on the $\{f_n\}$, in the statement of the lemma;
2. the fact that, for any value of i, the the following transformation of Φ_n results in Φ_{n+m-1}:
   a. replace 1 in the ith position of every term by $1 \otimes \cdots \otimes 1$ (m factors);
   b. replace φ in the ith position by Φ_m;
   c. replace ϵ in the ith position by $\epsilon \otimes \cdots \otimes \epsilon$ (m factors).

Claim 2: For all i Φ_i is *self-annihilating*. Consequently, the boundary map on the image of Φ_i is *injective*.

This follows from the Koszul convention and direct computation.

These claims imply that it suffices to show that $f_m(C) \circ_i (f_n(C)(c)(\sigma_1))(\sigma_2)$ and $f_{n+m-1}(C)(c)(\circledast(R(S_{i-1}) \otimes \underbrace{\mathfrak{T}_{1,\ldots,m,\ldots,1}}_{i^{\text{th}} \text{ position}}))(\sigma_1 \otimes \sigma_2)$, have the *same boundaries*. But this follows from the inductive hypothesis. $\square$

The remainder of this section will be spent defining the functors $\mathcal{C}(X)$ and $\mathfrak{C}(X)$, and developing some of their elementary properties. Both functors equip chain-complexes of X with coherent m-structures. The functor $\mathcal{C}(X)$ is defined over simplicial complexes and $\mathfrak{C}(X)$ is a singular version of $\mathcal{C}(X)$.

We begin with $\mathcal{C}(X)$. We define it to be a functor from the category of simplicial complexes to $\mathfrak{M}$ (see 3.10 on page 25). We define this functor to be free on the models $\{\Delta^p\}$, where Δ^p is the standard p-simplex. To (inductively) define the functor on Δ^p we first compute $\mathcal{C}(\partial\Delta^p)$ (where $\partial\Delta^p$ is the boundary of Δ^p), using the definition of $\mathcal{C}(*)$ on lower-dimensional simplices, and extend it to Δ^p using the fact that the latter is acyclic with a given contracting homotopy. We then use 4.1 to verify that the result is coherent.

The requirement that the m-structure be natural means that, when we have defined it on a k-simplex, we use that to compute that m-structure on the boundary of the standard $k+1$-simplex. Here:

1. $F_i([0, \ldots, m])$ = result of omitting i from this list.
2. The contracting homotopy is

$$(4.1) \qquad \mathfrak{s}([i_1, \ldots, i_k]) = \begin{cases} (-1)^k [i_1, \ldots, i_k, m] & \text{if } i_k \neq m, \text{and} \\ 0 & \text{otherwise.} \end{cases}$$

The augmentation is z, where $z([0]) = [m]$ and z of any other sequence $= 0$. The inductive definition is completed by defining $f_n([0, \ldots, m])([\ ]) = \delta^{n-1}$, where $\delta([0, \ldots, i]) = \sum_{j=0}^{i}[0, \ldots, j] \otimes [j, \ldots, i]$. The image of Φ_n on canonical basis elements of $R(S_n)$ is generated by tensor products of the form $[i_1, 1, \ldots, i_s, 1] \otimes \cdots \otimes [i_1, k, \ldots, i_t, k, m], [i_1, 1, \ldots, i_u, 1] \otimes \cdots \otimes [i_1, k, \ldots, i_v, k-1, m] \otimes [m]$, where the rightmost factor of length > 1 is guaranteed to have m in it, and all factors to the right of that are identically $[m]$.

The result of this procedure is a functor $\mathcal{C}(*)$ defined on simplices, that commutes with the face-operators (by fiat), in the sense that $\mathcal{C}(F_i\Delta^p) = \mathcal{C}(Z_i)\mathcal{C}(\Delta^{p-1})$, where $Z_i\colon \Delta^{p-1} \to \Delta^p$ is the map that includes Δ^{p-1} in Δ^p as the ith face. We can immediately extend this to all simplicial complexes. We also use it to define a coherent m-structure on the singular chain-complex of a topological space.

DEFINITION 4.2. Let X be a simplicial complex. For all $n \geq 0$ define $\mathcal{C}(X)_n$ to be freely generated by the n-simplices of X, with the m-structure induced by that of $\mathcal{C}(*)$ of these n-simplices. If X is a CW-complex, define $\mathfrak{C}(X)$ to be $\mathcal{C}(D(X))$, where $D(X)$ is the simplicial complex composed of the singular simplices of X.

Remarks.   4.2.1. In the case where X is a CW-complex, it is well-known that the topological realization of $D(X)$ is homotopy-equivalent to X.

4.2.2. It is not hard to see that the algebraically defined Steenrod operations (see 3.4 on page 23) for $\mathcal{C}(X)$ and $\mathfrak{C}(X)$ coincide with the geometrically defined Steenrod operations of X.

The functors $\mathcal{C}(*)$ and $\mathfrak{C}(*)$ clearly contain all of the information that can be derived from the chain-complex, as well as that contained in the coproduct. The underlying coproduct of $\mathcal{C}(*)$ and $C(*)$ is nothing but the *Alexander-Whitney diagonal map.*

The following results are clear:

PROPOSITION 4.3. *Let $f\colon X \to Y$ be a simplicial map of simplicial complexes. Then f induces a strict morphism of m-coalgebras: $\mathcal{C}(f)\colon \mathcal{C}(X) \to \mathcal{C}(Y)$. If $f\colon X \to Y$ is a map of CW-complexes, then f induces an equivalence of the m-coalgebras $\mathfrak{C}(f)\colon \mathfrak{C}(X) \to \mathfrak{C}(Y)$. It follows that $\mathcal{C}(*)$ and $\mathfrak{C}(*)$ are functors:*

$$\mathcal{C}(*)\colon(Simplicial\ complexes,\ simplicial\ maps) \to \mathfrak{M}_0$$

$$\mathfrak{C}(*)\colon(Topological\ spaces,\ continuous\ maps) \to \mathfrak{M}_0$$

REMARK. 4.3.1. Recall that $\mathfrak{M}_0$ is the category of coherent m-coalgebras whose morphisms are strict morphisms — see 3.8.1 on page 24. The fact that these functors produce *coherent* m-coalgebras follows from the fact that they are modeled upon simplices, and the chain-complexes of simplices have a natural structure of a coherent m-coalgebra.

PROPOSITION 4.4. *Let X be a polyhedron. Then there exists a natural equivalence of m-coalgebras $\mathcal{C}(X) \to \mathfrak{C}(X)$ in $\mathfrak{M}$.*

PROOF. This map results from regarding the inclusions of simplices of of X as singular simplices. In order to show that it is an equivalence in $\mathfrak{M}$, we must show that this map is the injection of a contraction of chain-complexes. This follows from the argument given in [13, §8.6] proving the equivalence of simplicial and singular chain-complexes. The equivalence is constructed there in two steps:

1. there is a map $u: C(X) \to D^V(|X|)$, where
   a. $C(X)$ is the simplicial chain-complex;
   b. V is an open cover of $|X|$;
   c. $D^V(|X|)$ is the singular chain-complex generated by singular simplices restricted to the open cover V;
2. $\nu: D^V(|X|) \to D(|X|)$, the map to the full singular chain-complex. In both cases, the maps induce strict morphisms of m-coalgebras, since they are induced by inclusions of geometric objects.

In case 1, Hilton and Wylie give an explicit inverse, $\gamma: D^V(|X|) \to C(X)$ to the map u and it is not hard to construct a nullhomotopy, ϕ, of $1 - u \circ \gamma$ with the required properties: if s is a singular k-simplex of X, we define $\phi(s)$ to be 0 if $s = u \circ \gamma(s)$ and to be to be a single singular simplex otherwise. If $u \circ \gamma(s) = 0$, we define $\gamma(s)$ to be the cone on s with respect to the barycenter — this is clearly degenerate. We want its boundary to be $(1 - u \circ \gamma)(s) - \phi(\partial s)$. Let one face of $\phi(s)$ be $u \circ \gamma(s)$; let the opposite vertex be the barycenter of s itself; and let the remaining faces of $\phi(s)$ be the result of gluing portions of the barycentric subdivision of s to corresponding simplices of $\phi(\partial s)$. This defines the mapping of the boundary of the standard $k+1$-simplex into $|X|$ — now extend this map to the entire $k+1$-simplex (such an extension is clearly possible) by taking a convex linear combination of the mapping on the boundary and using the embedding of X in $\mathbb{R}^\infty$ defined in [13, § 8.6]. This chain-homotopy has the required properties because:

1. it annihilates u since $\gamma \circ u = 1$;
2. γ annihilates it since $\gamma(\phi(s))$ will be degenerate — this makes use of the fact that γ is defined in terms of vertices of singular simplices;
3. it is self-annihilating since $\phi(s)$ is degenerate, where s is a simplex such that $\gamma \circ u(s) = 0$.

In case 2, the conclusion follows from the proof of 8.5.1 in [13]. □

The $\mathcal{C}(*)$ and $\mathfrak{C}(*)$-functors also satisfy a form of the Eilenberg-Zilber theorem:

THEOREM 4.5. *Let X and Y be simplicial complexes and let $g: \mathcal{C}(X) \otimes \mathcal{C}(Y) \to \mathcal{C}(X \times Y)$ be the unique natural map defined by the Eilenberg-Zilber theorem. Then g is the underlying map of a natural morphism of coherent m-structures, $(\iota, u, v, \varphi): \mathcal{C}(X) \otimes \mathcal{C}(Y) \to F(X, Y) \to \mathcal{C}(X \times Y)$. If either space is a point then this natural morphism is the identity map. The corresponding statement is true for CW complexes and the $\mathfrak{C}(*)$-functor.*

Remarks.   4.5.1. See [18] for a proof of the uniqueness of the map g. Note that the inverse map, $f: \mathcal{C}(X \times Y) \to \mathcal{C}(X) \otimes \mathcal{C}(Y)$ also preserves m-structures up to a natural chain-homotopy. The notation $(\iota, u, v, \varphi): \mathcal{C}(X) \otimes \mathcal{C}(Y) \to F(X, Y) \to \mathcal{C}(X \times Y)$ is defined in 3.11 on page 25.

4.5.2. Naturality of the morphism means that the contraction used to define the morphism is natural with respect to maps of X and Y.

4.5.3. This result provides some motivation for our definition of morphisms of m-coalgebras (given in 3.10 on page 25): we defined them in that way so that the maps in the Eilenberg-Zilber theorem would be morphisms.

PROOF. The complex $F(X, Y)$ is nothing but the pointed algebraic mapping cylinder of g^6. There clearly exists a contraction of this to $\mathcal{C}(X \times Y)$. We build a coherent m-structure on it via an application of 4.1 on page 28. Consider $F(\Delta^p, \Delta^q)$, where Δ^p and Δ^q are standard simplices of dimension p and q, respectively. Then the boundary map of $F(\Delta^p, \Delta^q)$ is given by

(4.2)
$$\begin{pmatrix} \partial_\otimes & -\downarrow & 0 \\ 0 & -\uparrow \circ \partial_\otimes \circ \downarrow & 0 \\ 0 & -g \circ \downarrow & \partial_\times \end{pmatrix} : C(\Delta^p) \otimes C(\Delta^q) \oplus \Sigma[C(\Delta^p) \otimes C(\Delta^q)] \oplus C(\Delta^{p+q} \times \Delta^{p+q})$$
$$\to C(\Delta^p) \otimes C(\Delta^q) \oplus \Sigma[C(\Delta^p) \otimes C(\Delta^q)] \oplus C(\Delta^{p+q} \times \Delta^{p+q})$$

Here $\partial_\otimes$ is the boundary of $C(\Delta^p) \otimes C(\Delta^q)$, $\partial_\times$ is that of $C(\Delta^{p+q} \times \Delta^{p+q})$, and $\uparrow$ and $\downarrow$ are, respectively, suspension and desuspension isomorphisms.

A contracting chain-homotopy for this is given by

(4.3) $\quad \Phi = \begin{pmatrix} \varphi_\otimes & \downarrow & 0 \\ 0 & -\uparrow \circ \varphi_\otimes \circ \downarrow & 0 \\ 0 & A & \varphi_\times \end{pmatrix} :$
$$C(\Delta^p) \otimes C(\Delta^q) \oplus \Sigma[C(\Delta^p) \otimes C(\Delta^q)] \oplus C(\Delta^{p+q} \times \Delta^{p+q})$$
$$\to C(\Delta^p) \otimes C(\Delta^q) \oplus \Sigma[C(\Delta^p) \otimes C(\Delta^q)] \oplus C(\Delta^{p+q} \times \Delta^{p+q})$$

Here $\varphi_\otimes \colon C(\Delta^p) \otimes C(\Delta^q) \to C(\Delta^p) \otimes C(\Delta^q)$ is a contracting homotopy for the tensor product. It is computed from the homotopy $\mathfrak{s}$ described in equation (4.1) on page 29, and $\varphi_\times \colon C(\Delta^{p+q} \times \Delta^{p+q}) \to \Delta^{p+q} \times \Delta^{p+q})$ is the corresponding contracting homotopy for pairs of simplices.

The map $A \colon \Sigma[C(\Delta^p) \otimes C(\Delta^q)] \to C(\Delta^{p+q} \times \Delta^{p+q})$ satisfies $g \circ \varphi_\otimes - \varphi_\times \circ g = A \circ \uparrow \circ \partial_\otimes - \partial_\times \circ A$. It is a chain-homotopy between the $g \circ \varphi_\otimes$ and $\varphi_\times \circ g$ — it measures the extent to which g maps $\varphi_\otimes$ into $\varphi_\times$. It exists (i.e. $g \circ \varphi_\otimes$ and $\varphi_\times \circ g$ are homotopic) because g is a chain-map and $\Phi(\Delta^p, \Delta^q)$ is acyclic. We can, consequently, compute Φ.

We will want Φ to be self-annihilating. This imposes the additional condition that $A \circ \uparrow \circ \varphi_\otimes \circ \downarrow = \varphi_\times \circ A$. A straightforward calculation shows that $\partial(A \circ \uparrow \circ \varphi_\otimes \circ \downarrow - \varphi_\times \circ A) = 0$, so that the condition is already satisfied modulo boundaries. We claim that it is possible to modify A so that this the condition will be satisfied exactly. We will do this by induction on dimension — clearly the condition is satisfied in the lowest-dimensional case since $A = 0$ there.

Suppose the condition is satisfied up to dimension k. The modifications of A will have to be carried out on the copy of A in the term $A \circ \uparrow \circ \varphi_\otimes \circ \downarrow$ since the other term has a lower-dimensional component of A. The modification will be possible if and only if $(A \circ \uparrow \circ \varphi_\otimes \circ \downarrow - \varphi_\times \circ A)| \ker(\uparrow \circ \varphi_\otimes \circ \downarrow) = 0$. But this kernel is equal

---

[6] This is the result of forming the ordinary algebraic mapping cylinder and collapsing basepoint $\times I$ to the basepoint.

to the image of $\uparrow \circ \varphi_\otimes \circ \downarrow$. Consequently, the modification is possible if and only if $(A \circ \uparrow \circ \varphi_\otimes \circ \downarrow - \varphi_\times \circ A) \circ \uparrow \circ \varphi_\otimes \circ \downarrow = \varphi_\times \circ A \circ \uparrow \circ \varphi_\otimes \circ \downarrow = 0$. But the inductive hypothesis implies that $A \circ \uparrow \circ \varphi_\otimes \circ \downarrow = \varphi_\times \circ A$ (since we have reduced it to one lower dimension), and the conclusion follows from the fact that $\varphi_\times$ is self-annihilating.

We now use F to inductively construct a coherent m-structure on $F(\Delta^p, \Delta^q)$ using 4.1 on page 28. We construct it to be a natural map — the inductive hypothesis implies that it already exists for all $F(\Delta^{p'}, \Delta^{q'})$, where either $p' < p$ and $q' = q$ or $q' < q$ and $p' = p$. In addition, we require the m-structure to coincide with those of the embedded $\mathcal{C}(\Delta^p) \otimes \mathcal{C}(\Delta^q)$ and $\mathcal{C}(\Delta^p \times \Delta^q)$. This is possible since F restricts to $\varphi_\otimes$ and $\varphi_\times$. $\square$

COROLLARY 4.6. *Let X be a simplicial H-space. Then $\mathcal{C}(X)$ is an m-Hopf algebra.*

CHAPTER 3

The bar and cobar constructions

1. $A(\infty)$-coalgebras

In this chapter we obtain formulas for the coproduct structure (and higher coproduct structure) of the cobar construction of a space in terms (essentially) of chain-level descriptions of Steenrod operations on the space. These are applied to determining the coproduct structure of the total space of fibrations.

In this section we will define and develop some of the properties of $A(\infty)$-coalgebras and $A(\infty)$-algebras. $A(\infty)$-coalgebras represent a *homotopy-invariant* version of co-associative coalgebras. Δ_2 corresponds to the coproduct and the higher coalgebra maps Δ_i, $i > 2$ measure deviations from co-associativity. These algebraic structures were discovered by Stasheff in [27]. Also see the discussion in the introduction of [17] and following 2.1 on page 38 in the present paper.

DEFINITION 1.1. A formal coalgebra U will be called a formal $A(\infty)$-coalgebra if it contains a set of elements $\{\Delta_i\}$, $i = 1, \ldots \infty$ where, $\mathrm{rank}(\Delta_i) = i$ and the $\dim(\Delta_i) = i - 2$, subject to the identity:

$$\sum_{k=1}^{n} \sum_{\lambda=0}^{n-k} (-1)^{k+\lambda+k\lambda} \Delta_k \circ_{n-k-\lambda+1} \Delta_{n-k+1} = 0$$

for all $n = 1, \ldots \infty$. In addition, we assume that the differential of U satisfies the condition $\partial \Delta_i = \sum_{j=1}^{i} \Delta_1 \circ_j \Delta_i + (-1)^{i+1} \Delta_i \circ_1 \Delta_1$.

Remarks. 1.1.1. The identity above implies that $(\Delta_1)^2 = 0$.

1.1.2. Note that it is possible to define a free formal $A(\infty)$-coalgebra — it has a $\mathbb{Z}$-basis such that all basis elements are composites of the $\{\Delta_i\}$, and the only identities they satisfy are the ones given above. In other words it the the free universal $\mathbb{Z}$-algebra on the symbols $\{\Delta_i\}$, modulo the ideal generated by the relation given above, and equipped with the differential given above. Here we define the differential of a composite via the usual convention: $\partial(a \circ_n b) = \partial(a) \circ_n b + (-1)^{\deg(a)} a \circ_n \partial(b)$, for pure elements a, and b.

We will denote this by $\mathfrak{F} = \{\mathfrak{F}_n\}$, where $\mathfrak{F}_n$ is the submodule generated by elements of rank n — it is essentially a universal model for an $A(\infty)$-coalgebra. A general formal $A(\infty)$-coalgebra may be regarded as the homomorphic image of $\mathfrak{F}$ in some formal coalgebra.

DEFINITION 1.2. If F is a formal coalgebra, then a *formal $A(\infty)$-coalgebra structure on* F will be defined to be a morphism of formal coalgebras:

$$\mathfrak{F} \to F$$

where $\mathfrak{F}$ is the free formal $A(\infty)$-coalgebra defined above.

With this in mind, we are in a position to define:

DEFINITION 1.3. An *actual $A(\infty)$-coalgebra* consists of:

1. a chain module C;
2. a sequence of chain-maps $g_n\colon C \to \mathrm{Hom}_{\mathbb{Z}}(\mathfrak{F}_n, C^n)$, with the property that:
   $g_1 = \partial\colon C \to C$, the differential of C (here we make use of the fact that $\mathfrak{F}_1 = \mathbb{Z}$ so $\mathrm{Hom}_{\mathbb{Z}}(\mathfrak{F}_1, C) = \Sigma C$);

Remarks.   1.3.1. We use what may seem to be an overly abstract definition of $A(\infty)$-coalgebras because we will exploit certain identities that exist in all such structures.

DEFINITION 1.4. Given two formal $A(\infty)$-coalgebras U, U' that are the images of $\mathfrak{F}$ in some formal coalgebra, Z, with generators $\{\Delta_i\}$, $\{\Delta_i'\}$, respectively, a *formal morphism* between these formal $A(\infty)$-coalgebras consists of elements $\{f_i\} \in Z$ of degree $i - 1$ such that the following equations are satisfied:

$$\sum_{j=1}^{q} \sum_{k_1+\cdots+k_j=q} (-1)^{\sum_{1 \leq \alpha < \beta \leq j}(k_\alpha+1)k_\beta} f_{k_1} \circ_1 \ldots f_{k_j} \circ_j \Delta_j$$

$$= \sum_{j=1}^{q} \sum_{\lambda=0}^{q-j}(-1)^{q+j+\lambda+j\lambda}\Delta_j' \circ_{q-\lambda-j+1} f_{q-j+1}$$

REMARK. 1.4.1. Given actual $A(\infty)$-coalgebras, C and C', modeled after formal $A(\infty)$-coalgebras U, U' as in 1.3, a formal morphism induces in like fashion, a sequence of maps $f_i'\colon C \to C'^n$. These maps define a morphism of $A(\infty)$-coalgebras as defined in [17, Définition 3.4]. Such a morphism can be characterized by the fact that it induces a DGA-algebra homomorphism of the respective cobar-constructions of C and C' — see [17, Définition 3.3] and [22].

PROPOSITION 1.5. *Let $\mathfrak{R}$ be an f-resolution whose composition maps are chain-maps. Let $z_2 \in (\mathfrak{R}_2)_0$ be an element mapping to $1 \in \mathbb{Z} = H_0(\mathfrak{R}_2)$ (see 3.1 on page 19). Then there exists a morphism $f(z_2)\colon \mathfrak{F} \to \mathfrak{R}$, sending Δ_2 to z_2, that is a morphism of formal coalgebras Any two such morphisms have the property that there exists a morphism (of formal $A(\infty)$-structures) between them whose 0 dimensional component is the identity element of $\mathfrak{R}$.*

Remarks. 1.5.1. Since z_2 is a cycle and the composition-operations of R are chain-maps, the $z_2 \circ_i z_2$ will also be cycles ($i = 1, 2$) — here $[z_2 \circ_i z_2]$ denotes the homology class of $z_2 \circ_i z_2$. These homology classes may be nontrivial since they are in dimension 0 and $\mathfrak{R}$ is only required to be acyclic above dimension 0. $\square$

1.5.2. This essentially implies that any weakly-coherent m-coalgebra has an associated $A(\infty)$-coalgebra structure — and that this structure is essentially unique. The map $\mathfrak{F} \to \mathfrak{R}$ induces a natural transformation of functors $\mathrm{Hom}_{\mathbb{Z} S_n}(\mathfrak{R}_n, *) \to \mathrm{Hom}_{\mathbb{Z} S_n}(\mathfrak{F}_n, *)$. Simply compose the structure morphism of the m-coalgebra with this natural transformation to get maps $g_n : C \to \mathrm{Hom}_{\mathbb{Z}}(\mathfrak{R}_n, C^n)$ defining an $A(\infty)$-coalgebra structure on C.

1.5.3. Note that this and 3.2 on page 19 provide a kind of partial solution to the (still open — see [17]) question of how one can form a natural tensor product of $A(\infty)$-coalgebras: if the $A(\infty)$-coalgebras in question are derived from weakly-coherent m-coalgebras the tensor-product is a weakly-coherent m-coalgebra that comes equipped with a natural $A(\infty)$-coalgebra. This does not solve the full problem, however, since the $A(\infty)$-coalgebra structure of the tensor product makes use of data not contained in the $A(\infty)$-structures of the factors. It is conceivable that there doesn't exist an intrinsic natural $A(\infty)$-coalgebra structure on the tensor product of two $A(\infty)$-coalgebras.

PROOF. We must show that the defining property of formal $A(\infty)$-coalgebras holds, at least on the image of f. We construct this map by induction on the generating set $\{\Delta_i\}$ of $\mathfrak{F}^+$. We begin by extending f to all compositions of Δ_2. Having extended f to the sub-formal $A(\infty)$-coalgebra generated by $\{\Delta_2, \ldots, \Delta_{n-1}\}$ we extend it to $\Delta_n - 1$ as follows: Compute

$$\bar{d}_n = \sum_{k=2}^{n} \sum_{\lambda=0}^{n-k} (-1)^{k+\lambda+k\lambda} \Delta_k \circ_{n-k-\lambda+1} \Delta_{n-k+1}$$

and $\bar{z}_n = f(\bar{d}_n)$.

CLAIM 1.5.1. $\bar{z}_n$ *is a cycle in* $\mathfrak{R}$. *This follows from the fact that the boundary of in* $\mathfrak{F}^+$ *is zero and the map constructed in the inductive hypothesis is a chain-map that preserves all of the composition operations. The acyclicity of* $\mathfrak{R}$ *implies that there exists an element* z_n *such that* $\partial z_n = \bar{z}_n$. *Set* $z_n = \varphi \bar{z}_n$, *where* φ *is the intrinsic contracting homotopy of* $\mathfrak{R}$ (*see 3.1 on page 19*) *and define* $f(\Delta_n) = z_n$ *and extend* f *to all possible composites with* $\{\Delta_2, \ldots, \Delta_n\}$

The second statement follows from the acyclicity of $\mathfrak{R}$, the freeness of $\mathfrak{F}^+$ (as a $\mathbb{Z}$-module) and the fact that we can separate off the cases where $j = 1$ in the formula in 1.4 on page 36 to inductively construct the morphism. $\square$

LEMMA 1.6. *Let* $(C, \{f[C]_n : C \to \mathrm{Hom}_{\mathbb{Z} S_n}(\mathfrak{R}[C]_n, C^n)\})$ *denote a weakly-coherent m-coalgebra. Then we have the following identity when evaluating the structure maps on an element,* $c \in C$:

$$\underbrace{1 \otimes \cdots \otimes \Delta_i \otimes \cdots \otimes 1}_{j^{th} \; position} = (-1)^{\dim(c)i} f[C]_{n+|\alpha|+i-1}(c) \circ (\Delta_i \circ_j *)$$

PROOF. This can be seen by performing the corresponding computation using the adjoint form of the structure maps, or by following a general rule of thumb that sliding one term to the right of another results in a sign of $(-1)^{\text{product of the degrees}}$ (the degree of Δ_i is $i - 2 \equiv i \pmod 2$). $\square$

2. The Cobar Construction

In this section $C = (C, \{\mathfrak{f}[C]_n : C \to \mathrm{Hom}_{\mathbb{Z} S_n}(\mathfrak{R}[C]_n, C^n)\})$ will denote a weakly coherent m-coalgebra with the property that $C_0 = \mathbb{Z}$, and for some value of $k > 1$, $C_i = 0$ for $i < k$. We will essentially show that the cobar construction is well-defined for C, and that this construction itself comes equipped with a natural structure as a weakly coherent m-coalgebra. We begin by recalling the definition of the cobar construction of an $A(\infty)$-coalgebra (see [17, Définition 3.2]

DEFINITION 2.1. If C is some $A(\infty)$-coalgebra with $C_0 = \mathbb{Z}$, and $C_i = 0$ for $0 < i < k$, the *cobar construction*, $\mathcal{F}(C)$, is defined to be $T(\Sigma^{-1} C^+)$, with boundary map on $\Sigma^{-1} C^+$ given by

$$\partial_{\text{cobar}} = -\sum_{i=1}^{\infty} \underbrace{\downarrow \otimes \ldots \downarrow}_{i \text{ copies}} \circ \Delta_i \circ \uparrow : \Sigma^{-1} C \to T(\Sigma^{-1} C)$$

Here $\uparrow : \Sigma^{-1} C^+ \to C^+$ sends an element its suspension, $\downarrow$ is the inverse of $\uparrow$. This definition is extended to all of $T(\Sigma^{-1} C^+)$ by the requirement that it form a DGA-algebra.

Remarks.   2.1.1. The cobar construction was originally defined for DGA-coalgebras by Adams in [1]. He showed that the cobar construction of the singular chain-complex of a pointed simply-connected topological space (with its geometrically-defined) coalgebra structure, is essentially the chain complex of its *loop space*.

2.1.2. Here $\{\Delta_i\}$ represents the canonical $A(\infty)$-coalgebra structure

2.1.3. Many of the algebraic constructs used in topology that are based upon co-associative coalgebras can be re-defined in terms of $A(\infty)$-coalgebras. This is, in fact, the primary motivation for working with $A(\infty)$-coalgebras. The point is that co-associativity of a coalgebra is not a homotopy invariant property while the property of being an $A(\infty)$-coalgebras is. Here is an example of such a construct that will be used in the sequel:

DEFINITION 2.2. Let C be an $A(\infty)$-coalgebra; let A be a DGA-algebra and let $x : C \to A$ is a map of degree -1. This map will be called a *twisting cochain* if

$$\partial_A \circ x + \sum_{i=1}^{\infty} \mu^{i-1} \circ x^i \circ \Delta_i = 0$$

In this case the map $\partial_x : C \otimes A \to C \otimes A$ is a *differential*, where

$$\partial_x = 1 \otimes \partial_A + \sum_{i=1}^{\infty} (1 \otimes \mu^{i-1}) \circ (1 \otimes x^{i-1} \otimes 1) \circ (\Delta_i \otimes 1)$$

Equipped with this differential, $C \otimes A$ will be denoted $C \otimes_\xi A$ and called the *twisted tensor product* with respect to the twisting cochain ξ. The $A(\infty)$-coalgebra C will be called the *base* of the twisted tensor product and A will be called its *fiber*.

Remarks. 2.2.1. At first glance it might appear that the boundary of C doesn't enter into the definition of a twisted tensor product. This is a mistaken impression — we use the fact that $\Delta_1 = \partial_C$ in an $A(\infty)$-coalgebra, so the formulas above could be re-written:

$$x \circ \partial_C + \partial_A \circ x + \sum_{i=2}^{\infty} \mu^{i-1} \circ x^i \circ \Delta_i = 0$$

and

$$\partial_x = \partial_C \otimes 1 + 1 \otimes \partial_A + \sum_{i=2}^{\infty} (1 \otimes \mu^{i-1}) \circ (1 \otimes x^{i-1} \otimes 1) \circ (\Delta_i \otimes 1)$$

2.2.2. We can also define a *dual* concept — a twisted tensor product formed with respect to a DGA-coalgebra and an $A(\infty)$-algebra. See 2.5 on page 40.

2.2.3. In actual applications C will almost always be a weakly-coherent m-coalgebra and we will use the canonical associated $A(\infty)$-coalgebra in forming twisted tensor products. In the situation described here it is not too difficult to define a morphism of $A(\infty)$-coalgebras $b(x): C \to \bar{\mathcal{B}}(A)$ — see [17]. Essentially, define

$$\bar{\Delta} = \sum_{i=2}^{\infty} (1 \otimes \mu^{i-2}) \circ (1 \otimes \xi^{i-1}) \circ \Delta_i : C \to C \otimes A$$

and

$$\bar{\Delta}^i = \begin{cases} 1: C \to C & \text{if } i = 1 \\ \dots (\bar{\Delta} \otimes 1 \otimes \dots \otimes 1) \circ \dots \circ (\bar{\Delta} \otimes 1) \circ \bar{\Delta} & \text{if } i > 1 \end{cases}$$

$(i - 1$ copies of $\bar{\Delta}$, when $i > 1$). Now define

$$b(x) = \epsilon + \sum_{i=1}^{\infty} T_i \circ \underbrace{\uparrow \otimes \cdots \otimes \uparrow}_{i \text{ copies}} (x \otimes 1 \otimes \cdots \otimes 1) \circ \bar{\Delta}^i$$

Here T_i is the map that identifies $\Sigma A \otimes \cdots \otimes \Sigma A$ with $[A|\dots|A] \subset \bar{\mathcal{B}}(A)$.

2.2.4. In more generality, if M is a left DGA A-module, and $C \otimes_x A$ is as defined above, then $(C \otimes_x A) \otimes_A M = C \otimes_x M$, is also called a twisted tensor product.

2.2.5. Twisting cochains were originally defined by Brown in [3]. He proved that the chain-complex of the total space of a fibration was chain-homotopy equivalent to a suitable twisted tensor product.

DEFINITION 2.3. Let C be an $A(\infty)$-coalgebra and let $\mathcal{F}(C)$ be the associated cobar construction (defined above in 2.1 on page 38). The map:

$$C \xrightarrow{\ell} \mathcal{F}(C)$$

is defined to send $c \in C$ to $\downarrow(c) \in \Sigma^{-1}C \subset \mathcal{F}(C)$.

This map is a twisting cochain, and the twisted tensor product $C \otimes_\ell \mathcal{F}(C)$ is acyclic. This is called the *canonical acyclic twisted tensor product* of C.

REMARK. 2.3.1. See § 3 of [17] for a proof of the fact that ℓ is a twisting cochain, and that $C \otimes_\ell \mathcal{F}(C)$. The statement that ℓ is a twisting cochain is synonymous with the statement that the differential of $\mathcal{F}(C)$ is self-annihilating.

In order to proceed further we will need the *dual* to an $A(\infty)$-coalgebra, namely an $A(\infty)$-algebra, and some associated concepts:

DEFINITION 2.4. Let A be a chain-complex. Then A will be called an $A(\infty)$-algebra if it is equipped with maps $\bar{\mu}_i \colon A^i \to A$ of degree $i - 2$ satisfying the identity

$$\sum_{k=1}^{n} \sum_{\lambda=1}^{n-k} (-1)^{k+\lambda+k\lambda} \bar{\mu}_{n-k+1} \circ (1^\lambda \otimes \bar{\mu}_k \otimes 1^{n-k-\lambda}) = 0$$

where $\bar{\mu}_1 = \partial \colon A \to A$;

Given an $A(\infty)$-algebra, $(A, \{\bar{\mu}_i\})$, the *associated bar construction* is defined to be the free DGA-coalgebra $T(\Sigma A)$, with differential defined on $[a_1|\ldots|a_k]$, with $a_j \in A$, to be

$$\uparrow(\circ \bar{\mu}_1 \circ \downarrow \otimes 1 \otimes 1 + \cdots + 1 \otimes \cdots \otimes \uparrow \circ \bar{\mu}_1 \circ \downarrow + (\uparrow \otimes \uparrow) \circ \bar{\mu}_2 \circ \downarrow \otimes \cdots \otimes 1$$
$$+ \cdots + 1 \otimes \cdots \otimes \uparrow \otimes \uparrow) \circ \bar{\mu}_2 \circ \downarrow + \cdots + (\uparrow \otimes \cdots \otimes \uparrow) \circ \bar{\mu}_k \circ \downarrow$$

Remarks.   2.4.1. $A(\infty)$-algebras were first developed by Stasheff in [27] as a way of algebraically describing H-spaces whose product operation was not associative.

2.4.2. Bar constructions were first defined by Eilenberg and MacLane in [7]. They showed that the n-fold iterated bar construction of the ring $\mathbb{Z}\pi$ is chain-homotopy equivalent to the chain-complex of the Eilenberg-MacLane space $K(\pi, n)$.

2.4.3. The defining identity for an $A(\infty)$-algebra turns out to be equivalent to the statement that the differential on the associated bar construction is self-annihilating — see [17, § 3].

We can define twisted tensor products in terms of $A(\infty)$-algebras rather than $A(\infty)$-coalgebras. The resulting definition is, in some sense, dual to that of 2.2 above:

DEFINITION 2.5. Let C be a DGA-coalgebra and let $(A, \{\bar{\mu}_i\})$ be an $A(\infty)$-algebra.

1. A map $x \colon C \to A$, of degree -1 will be called a twisting cochain if it satisfies the condition

$$x \circ \partial_C + \sum_{i=1}^{\infty} \bar{\mu}_i \circ \underbrace{x \otimes \cdots \otimes x}_{i \text{ factors}} \circ \Delta^{i-1} = 0$$

2. Given a twisting cochain, x, the map

$$b(x) = \epsilon + \sum_{i=1}^{\infty} T_i \circ \underbrace{x \otimes \cdots \otimes x}_{i \text{ factors}} \circ \Delta^{i-1} \colon C \to \bar{\mathcal{B}}(A)$$

is a homomorphism of DGA-coalgebras. As in remark 2.2.3 on page 39, T_i is the map that identifies $\Sigma A \otimes \cdots \otimes \Sigma A$ with $[A|\ldots|A] \subset \bar{\mathcal{B}}(A)$.

3. Given a twisting cochain, x, the map

$$\partial_x = 1 \otimes \partial_C + \sum_{i=1}^{\infty} (\bar{\mu}_i \otimes 1) \circ 1 \otimes \underbrace{x \otimes \cdots \otimes x}_{i-1 \text{ factors}} \otimes 1 \circ \Delta^{i-1} \otimes 1 \colon A \otimes C \to A \otimes C$$

is self-annihilating.

The chain-complex equipped $A \otimes C$ with the differential ∂_x will be called the *a-twisted tensor product* and will be denoted $A \bigstar_{x,\{\bar{\mu}_i\}} C$.

Remarks. 2.5.1. As in the previous definition of a twisted tensor product, we have incorporated the differential of one of the chain-complexes (namely A) via an $A(\infty)$-structure. Here $\bar{\mu}_1 = \partial_A$.

2.5.2. Both of the definitions of twisted tensor product in this section are essentially standard. See [17, § 3] for these definitions and the proof that a map $b \colon C \to \bar{\mathcal{B}}(A)$ is a homomorphism of DGA-coalgebras if and only if it is of the form $b(x)$ for some twisting cochain x.

Note that $x(C_0) = 0$ so that the 0-dimensional components of the coproduct, Δ, have no effect on the formula for $b(x)$ or the definition of a twisting cochain.

2.5.3. Note that these twisted tensor products are written as fiber×base rather than base×fiber. These twisted tensor products are used first on page 77 and the surrounding discussion.

2.5.4. If A is a DGA-algebra, the transposition map defines a *canonical isomorphism* of chain-complexes $T \colon A \bigstar_{x,\{\bar{\mu}_i\}} C \cong C \otimes_x A$, where C' is the DGA coalgebra with the same underlying chain-complex as C, but whose coproduct $\Delta_{C'} = T \circ \Delta_C$ and Δ_C is the coproduct of C. Here is the twisted tensor product formed with respect to the coalgebra structure of C' and the $A(\infty)$-algebra structure of A.

Note that $x(C_0) = 0$ so that the 0-dimensional components of the coproduct, Δ, have no effect on the formula for $b(x)$ or the definition of a twisting cochain.

2.5.5. As in remark 2.2.4 on page 39, we can generalize this definition of twisted tensor product somewhat. Let M be a left $A(\infty)$-module over A, i.e. suppose there exist maps $\{\hat{a}_k\}$ of degree $k - 2$, $\hat{a}_k \colon A \otimes \cdots \otimes A \otimes M \to M$ (k-factors in all), otherwise satisfying the same identities as an $A(\infty)$-product. This means that the equation in 2.4 on page 40 is satisfied with the rightmost copy of $\bar{\mu}_k$ replaced by $\hat{a}_k$. In this case we can define a twisted tensor product $C \otimes_{\hat{a}_k} M$.

It is possible to define morphisms of $A(\infty)$-algebras in a manner that is entirely dual to the definition of morphisms of $A(\infty)$-coalgebras. We get:

DEFINITION 2.6. Given two $A(\infty)$-algebras $(A, \{\bar{\mu}_i\})$, $(A', \{m'_i\})$, a *morphism* from

A to A' is a family of maps $\{f_i\}$, where f_i is of degree $i-1$, satisfying:

$$\sum_{j=1}^{q} \sum_{k_1+\cdots+k_j=q} (-1)^{\sum_{1\leq\alpha<\beta\leq j}(k_\alpha+1)k_\beta} m'_j \circ (f_{k_1} \otimes \cdots \otimes f_{k_j})$$

$$= \sum_{j=1}^{q} \sum_{\lambda=0}^{q-j} (-1)^{q+j+\lambda+j\lambda} f_{q-j+1} \circ (1_\lambda \otimes m_j \otimes 1_{q-j-\lambda})$$

REMARK. 2.6.1. See [17, § 3]. Morphisms of $A(\infty)$-algebras have the property that they induce DGA-coalgebra maps of the associated bar-constructions — this can, in fact, be taken as their definition.

DEFINITION 2.7. Let $\mathfrak{R}$ be an f-resolution (defined in 3.1 on page 19) whose composition-maps are chain-maps. Then $Z'_{n,m}(\mathfrak{R})$ is defined to be

$$\bigoplus_{\substack{\{\alpha_1,\ldots,\alpha_n\} \\ |\alpha|\leq m}} \Sigma^{-|\alpha|}\mathfrak{R}_{|\alpha|}$$

and $Z_{n,m}(\mathfrak{R})$ is defined to be the result of truncating $Z'_{n,m}(\mathfrak{R})$ in dimension -1. Here the summation is taken over all possible length-n sequences of integers $\{\alpha_1,\ldots,\alpha_n\}$ and $|\alpha|$ denotes the sum of the α_i. We will also define the projection-maps

$$p'_{m',m}\colon Z'_{n,m'}(\mathfrak{R}) \to Z'_{n,m}(\mathfrak{R})$$
$$p_{m',m}\colon Z_{n,m'}(\mathfrak{R}) \to Z_{n,m}(\mathfrak{R})$$

Remarks.   2.7.1. The summation is taken over all possible sequences of length n that satisfy the condition on the total.

2.7.2. We will sometimes want to consider individual summands in $Z'_{n,m}(\mathfrak{R})$. We will denote the summand $\Sigma^{-|\alpha|}\mathfrak{R}_{|\alpha|}$ corresponding to the sequence $\alpha = (\alpha_1,\ldots,\alpha_n)$ by $\Sigma^{-|\alpha|}\mathfrak{R}_\alpha$ or $\Sigma^{-|\alpha|}\mathfrak{R}_{\alpha_1,\ldots,\alpha_n}$. With this in mind $Z_{n,m}(\mathfrak{R})$ is equipped with an action of $\mathbb{Z}S_n$ defined as follows: if $a \in \Sigma^{-|\alpha|}\mathfrak{R}_\alpha$, $\sigma \in S_n$ then $\sigma \cdot a = \text{Parity}(\mathrm{T}_{\alpha_1,\ldots,\alpha_n}(\sigma))\mathrm{T}_{\alpha_1,\ldots,\alpha_n}(\sigma) \cdot a$, where the product on the right-hand side is taken in $\Sigma^{-|\alpha|}\mathfrak{R}_{|\alpha|}$ and the target is regarded as being an element of $\Sigma^{-|\alpha|}\mathfrak{R}_{\sigma^{-1}\{\alpha_1,\ldots,\alpha_n\}}$. Here $\sigma^{-1}\{\alpha_1,\ldots,\alpha_n\} = \{\alpha_{\sigma(1)},\ldots,\alpha_{\sigma(n)}\}$ is the result of permuting the elements of α via σ^{-1}. Note that we are defining the action of S_n to permute the summands of $Z'_{n,m}(\mathfrak{R})$, as well as twisting it by the parity of the permutations.

DEFINITION 2.8. Let $\mathfrak{R}$ be an f-resolution (defined in 3.1 on page 19). Suppose, in addition, that $\mathfrak{R}$ is equipped with a formal $A(\infty)$-coalgebra structure $\{\Delta_i\}$ — see 1.2 on page 36. Define $\hat{\mathfrak{T}}\colon Z'_{n,m}(\mathfrak{R}) \to Z'_{n,m}(\mathfrak{R})$ by

$$(2.1)\quad \hat{\mathfrak{T}}|\Sigma^{-|\alpha|}\mathfrak{R}_\alpha = p_{\infty,m} \circ \left\{ \sum_{i=1}^{|\alpha|} \sum_{j=1}^{\infty} (-1)^{i+|\alpha|j+ij} \downarrow^{|\alpha|+j-1} \circ(\Delta_j \circ_i *) \circ \uparrow^{|\alpha|} \right\}$$

— this map will be of degree -1.

Remarks. 2.8.1. Note that all f-resolutions satisfying the hypotheses above possess a formal $A(\infty)$-structure — see 1.5 on page 36

2.8.2. It is necessary to say something about precisely where the image of $\hat{\mathfrak{T}}$ lies. This is due to the fact that we have introduced some extra structure into $\Sigma^{-|\alpha|}\mathfrak{R}_{|\alpha|}$ — namely that of the sequence $\alpha = \{k_1, \ldots, k_n\}$.

DEFINITION 2.9. We assume that the image of $\downarrow^{|\alpha|+j-1} \circ (\Delta_j \circ_i *) \circ \uparrow^{|\alpha|}$ lies in $\Sigma^{-|\alpha(i,j)|}\mathfrak{R}_{\alpha(i,j)}$, where $\alpha(i,j)$ is the sequence $\{k_1, \ldots, k_{r(i)} + j - 1, \ldots, k_n\}$, and $r(i)$ is the *smallest* value of t such that $i < \sum_{l=1}^{t} \alpha_t$. Note that $|\alpha(i,j)| = |\alpha| + j - 1$. It is not hard to see that $\hat{\mathfrak{T}}$ is $\mathbb{Z}S_n$-linear, where the action of S_n is defined in remark 2.7.2. This observation makes use of 2 of A.2 appendix A (page 99).

We will need some notation in the sequel:

DEFINITION 2.10. The notation $\downarrow_k$ will be used to denote

$$\underbrace{\downarrow \otimes \cdots \otimes \downarrow}_{k \text{ copies}}$$

LEMMA 2.11. *Let* $C = (C, \{\mathfrak{f}[C]_n : C \rightarrow \mathrm{Hom}_{\mathbb{Z}S_n}(\mathfrak{R}[C]_n, C^n)\})$ *be a weakly-coherent m-coalgebra with associated $A(\infty)$-coalgebra structure $\{\Delta_i\}$ and let α be a sequence of nonnegative integers of length n. Then the diagram in figure 3.3.1 on page 44 commutes.*

REMARK. 2.11.1. Recall that $\widetilde{\mathfrak{f}[C]_n} : \mathfrak{R}[C]_n \otimes C \rightarrow C^n$ are the adjoints of the coordinate maps $\mathfrak{f}[C]_n : C \rightarrow \mathrm{Hom}_{\mathbb{Z}S_n}(\mathfrak{R}[C]_n, C^n)$. The commutativity of the diagrams follows by a repeated application of the Koszul Convention: we start with

$$1 \otimes \cdots \otimes \downarrow_j \circ \Delta_j \circ \uparrow \otimes \cdots \otimes 1 \circ \downarrow \otimes \cdots \otimes \downarrow \circ \mathfrak{f}[\widetilde{C}]_{|\alpha|} \circ \uparrow^{|\alpha|}$$

Now we push $\downarrow_j \circ \Delta_j \circ \uparrow$ (which is of degree -1) past $i - 1$ copies of $\downarrow$, producing $(-1)^{i-1} \downarrow \otimes \cdots \otimes \downarrow_j \circ \Delta_j \otimes \cdots \otimes \downarrow \circ \tilde{f}_{|\alpha|} \circ \uparrow^{|\alpha|}$. The remaining exponents of -1 occur as the result of pushing Δ_j (which is of degree $j - 2$) past the remaining $|\alpha| - i$ copies of $\downarrow$.

PROPOSITION 2.12. *Let $\mathfrak{R}$ be an f-resolution (defined in 3.1 on page 19). Suppose, in addition, that $\mathfrak{R}$ is equipped with a formal $A(\infty)$-coalgebra structure $\{\Delta_i\}$ — see 1.2 on page 36. The map, $\hat{\mathfrak{T}}$, defined in 2.8 on page 42, has the property that $\hat{\mathfrak{T}}^2 = 0$.*

Remarks. 2.12.1. This implies that $Z'_{n,m}(\mathfrak{R})$ and $Z_{n,m}(\mathfrak{R})$, equipped with this differential, constitute *chain-complexes*.

Diagram 3.3.1 on page 44 implies that the result of evaluating $\hat{\mathfrak{T}}$ on $\mathcal{F}(C)$ is a chain-map — it coincides with the differential of $\mathcal{F}(C)$ (see 2.1 on page 38). We want to show that $\hat{\mathfrak{T}}^2 = 0$ holds, as an *identity* in $Z'_{n,m}(\mathfrak{R})$.

2.12.2. It is not difficult to see that varying the canonical $A(\infty)$-coalgebra structure of $\mathfrak{R}$ by a formal isomorphism as in remark 1.4.1 on page 36 varys $Z'_{n,m}(\mathfrak{R})$ and $Z_{n,m}(\mathfrak{R})$ by an isomorphism of chain-complexes that is the identity map on the chain-level.

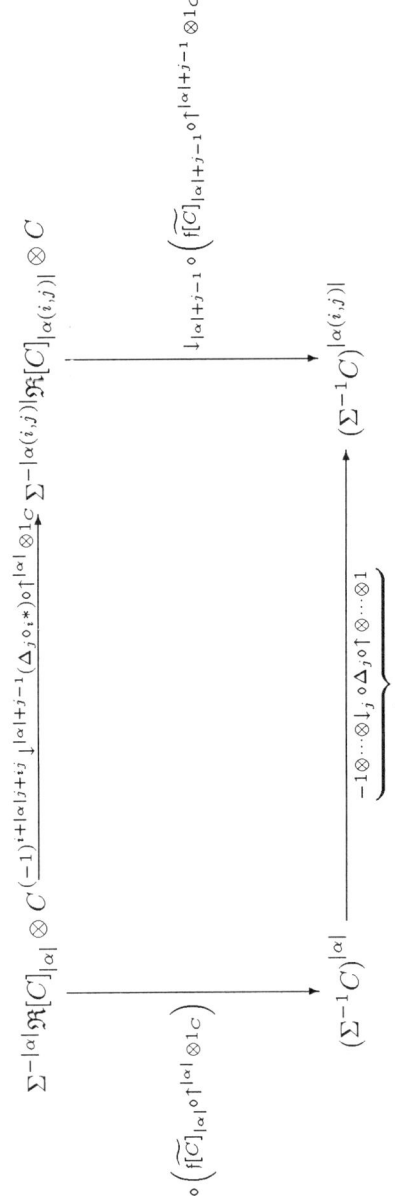

Figure 3.3.1. Correspondence between the Δ_i (as elements of $\mathfrak{R}[C]_n$) and the $A(\infty)$-structure of C

PROOF. This is proved in appendix C on page 106.   □

PROPOSITION 2.13. *Let $\mathfrak{R}$ be an f-resolution (defined in 3.1 on page 19) and suppose that $\mathfrak{R}$ is equipped with a formal $A(\infty)$-coalgebra structure $\{\Delta_i\}$ — see 1.2 on page 36. Then the chain-complexes $\{Z'_{n,m}(\mathfrak{R})\}$, defined in 2.8 on page 42, are acyclic for all values of n and in all dimensions. Consequently, the chain-complexes $Z_{n,m}(\mathfrak{R})$ are acyclic above dimension -1.*

PROOF. This follows by induction on m. Clearly, $Z'_{n,1}(\mathfrak{R})$ is acyclic in positive dimensions. $Z'_{n,i+1}(\mathfrak{R})$ is just an extension of $Z'_{n,i}(\mathfrak{R})$ by chain complexes that are acyclic above dimension $-(i+1)$. We claim that the connecting homomorphism of the extension kills off this additional homology in dimension $-(i+1)$. It suffices to consider the induced maps in homology of the boundary of $Z'_{n,i+1}(\mathfrak{R})$ — see 2.8 on page 42. The only significant contribution will be from Δ_0 (the ordinary coproduct) since it is a chain-map. It is not hard to see that Δ_0 will induce an injective map of all of the nontrivial homology of $Z'_{n,i}(\mathfrak{R})$ to the new chain-complexes of degree $-(i+1)$. Here, we are making use of the fact that the nontrivial homology of each of these chain-complexes is $\mathbb{Z}$, concentrated in a single dimension. Consequently, the nontrivial homology of $Z'_{n,i+1}(\mathfrak{R})$ will lie entirely in dimension $-(i+1)$, and $Z'_{n,i+1}(\mathfrak{R})$ will have no nontrivial homology.   □

DEFINITION 2.14. Let $\beta_1, \ldots, \beta_k$ be k sequences of nonnegative integers, where each sequence is of length n. Define:

1. $\mathcal{Z}\{\beta_1, \ldots, \beta_k\}$ to be the signed shuffle-permutation that shuffles the sequence $\{1, \ldots, q\} = \{\beta_{1,1}, \ldots, \beta_{1,n}, \ldots, \beta_{k,1}, \ldots, \beta_{k,n}\}$, where $q = \sum_{i=1}^{k} |\beta_i|$, into $\{\beta_{1,1}, \ldots, b_{k,1}, \ldots, \beta_{1,n}, \ldots, \beta_{k,n}\}$. Here each term $\beta_{i,j}$, itself, represents a sequence of $\beta_{i,j}$, consecutive integers. It follows that $\mathcal{Z}\{\beta_1, \ldots, \beta_k\} \in R(S_q)$, and the sign is just the parity of the permutation.
2. $V(\beta_1, \ldots, \beta_k)$ to be the corresponding shuffle-map of C^q.

Remarks. 2.14.1. Note that $V(\beta_1, \ldots, \beta_k)$ is a map of a tensor product of copies of C, while $\mathcal{Z}\{\beta_1, \ldots, \beta_k\}$ is an element of $R(S_q)$. By definition, $\mathcal{Z}\{\alpha\}$ (i.e., a single sequence) = the identity map.

2.14.2. Given this definition and an $A(\infty)$-coalgebra structure on $\mathfrak{R}$ (for instance if it is an f-resolution) — see 1.2 on page 36, we can define an $A(\infty)$-algebra structure on $Z_{n,m}(\mathfrak{R}[C])$: We first must define a k-linear product $t_k \colon \mathfrak{R}[C]_{i_1} \otimes \cdots \otimes \mathfrak{R}[C]_{i_k} \to \mathfrak{R}[C]_m$, of degree $k-2$ by $t_k(r_1 \otimes \cdots \otimes r_k) = (-1)^{k \sum_{i=1}^{k} \dim(r_i)} r_1 \circ_1 \ldots r_k \circ_k \Delta_k$.

If $\mathfrak{R} = \mathfrak{R}[C]$, where $C = (C, \{\mathfrak{f}[C]_n \colon C \to \mathrm{Hom}_{\mathbb{Z}S_n}(\mathfrak{R}[C]_n, C^n)\})$ is a weakly coherent m-coalgebra, then the basic property that the $\{t_k\}$ have is expressed by the fact that the diagram in figure 3.3.2 (on page 46) commutes.

where the map $\bigotimes_{j=1}^{k} \mathrm{Hom}_{\mathbb{Z}}(\mathfrak{R}[C]_{i_j}, C^{i_j}) \to \mathrm{Hom}_{\mathbb{Z}}(\bigotimes_{j=1}^{k} \mathfrak{R}[C]_{i_j}, C^v)$ sends tensor products of maps to maps of the tensor products of their domains and ranges.

DEFINITION 2.15. If $(C, \{\mathfrak{f}[C]_n \colon C \to \mathrm{Hom}_{\mathbb{Z}S_n}(\mathfrak{R}[C]_n, C^n)\})$ is a weakly coherent m-coalgebra, the products $\{t_k\}$ give rise to a natural k-linear product on $Z_{n,m}(\mathfrak{R}[C])$,

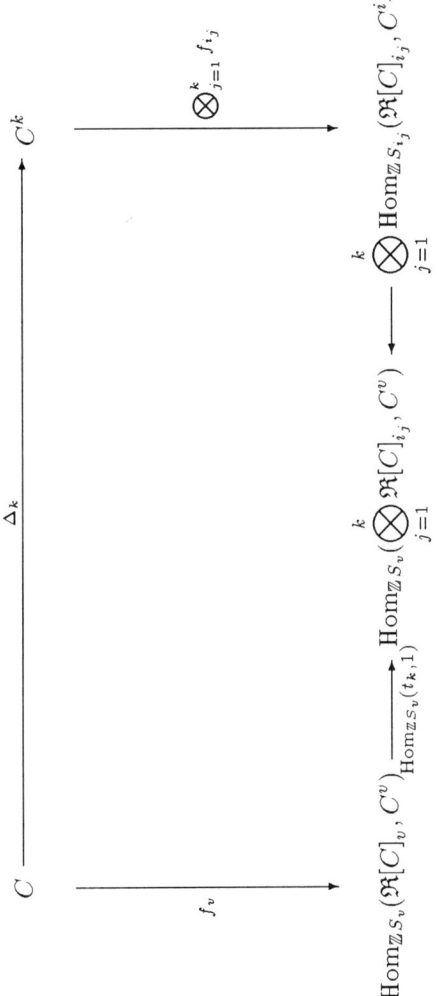

Figure 3.3.2.

of degree $k - 2$:

$$\bar{\mu}_k = (-1)^{k|\alpha|} p_{\infty,m} \circ \downarrow^{|\alpha|} \mathcal{Z}\{\beta_1, \ldots, \beta_k\} \circ t_k \circ \uparrow^{|\beta_1|} \otimes \cdots \otimes \uparrow^{|\beta_k|}$$

$$: \Sigma^{-|\beta_1|} \mathfrak{R}[C]_{\beta_1} \otimes \cdots \otimes \Sigma^{-|\beta_k|} \mathfrak{R}[C]_{\beta_k}$$

$$\rightarrow \Sigma^{-|\alpha|} \mathfrak{R}[C]_{\alpha}$$

where $\alpha = \sum_{i=1}^{k} \beta_i$ and the summation of sequences is taken elementwise. Here the $\{\beta_i\}$ are each sequences of nonnegative integers of length n. This product operation is $\mathbb{Z}S_n$-linear, where the action of S_n is defined in remark 2.7.2 on page 42. This is due to the fact that the maps (regarded as permutations) preserve the blocks $\{\beta_{i,1}, \ldots, \beta_{i,n}\}$ (see 2.7 on page 14).

LEMMA 2.16. *Let* $(C, \{\mathfrak{f}[C]_n : C \rightarrow \mathrm{Hom}_{\mathbb{Z}S_n}(\mathfrak{R}[C]_n, C^n)\})$ *be a weakly coherent m-coalgebra. Let* $\{\beta_i\}$, $i = 1, \ldots, k$ *each be sequences of nonnegative integers of length n and define evaluation-maps* $e(\beta_1, \ldots, \beta_k) : \Sigma^{-|\beta_1|} \mathfrak{R}[C]_{|\beta_1|} \otimes \cdots \otimes \Sigma^{-|\beta_k|} \mathfrak{R}[C]_{|\beta_k|} \otimes C^k \rightarrow$ $(\Sigma^{-1} C)^{|\alpha|} = \mathcal{F}(C)^n$ *via*

$$e(\beta_1, \ldots, \beta_k) = \downarrow_{|\alpha|} \circ V(\beta_1, \ldots, \beta_k) \circ$$

$$\left(\tilde{f}_{\beta_1} \otimes \cdots \otimes \tilde{f}_{\beta_k} \right) \circ V \circ \uparrow^{|\beta_1|} \otimes \cdots \otimes \uparrow^{|\beta_k|} \otimes 1 \otimes \cdots \otimes 1$$

— note that α is a sequence whose elements are sums of corresponding elements f the β_i.

Here V is the shuffle-map that shuffles the factors of $\Sigma^{-|\beta_1|} \mathfrak{R}[C]_{|\beta_1|} \otimes \cdots \otimes$ $\Sigma^{-|\beta_k|} \mathfrak{R}[C]_{|\beta_k|} \otimes C^k$ to get $\Sigma^{-|\beta_1|} \mathfrak{R}[C]_{|\beta_1|} \otimes C \otimes \cdots \otimes \Sigma^{-|\beta_k|} \mathfrak{R}[C]_{|\beta_k|} \otimes C$, and the $\mathfrak{f}[\tilde{C}]_i : \mathfrak{R}[C]_i \otimes C \rightarrow C^i$ are the adjoint maps associated with the structure maps $\mathfrak{f}[C]_i : C \rightarrow \mathrm{Hom}_{\mathbb{Z}S_i}(\mathfrak{R}[C]_i, C^i)$. Then:

$$e(\beta_1, \ldots, \beta_k) \circ 1 \otimes \cdots \otimes 1 \otimes \Delta_k = e(\alpha) \circ \bar{\mu}_k \otimes 1$$

where $\alpha = \sum_{i=1}^{k} \beta_i$ as in 2.15 on page 45

Remarks. 2.16.1. See 2.10 on page 43 for the notation $\downarrow_n$.

2.16.2. It is not hard to see that the $\{e(\beta_1, \ldots, \beta_k)\}$ are chain-maps $\Sigma^{-|\beta_1|} \mathfrak{R}[C]_{|\beta_1|} \otimes$ $\cdots \otimes \Sigma^{-|\beta_k|} \mathfrak{R}[C]_{|\beta_k|} \otimes C^k \rightarrow (\Sigma^{-1} C)^{|\alpha|}$. A diagram-chase and 2.11 show that, if we regard the $\Sigma^{-|\beta_j|} \mathfrak{R}[C]_{\beta_j}$ as summands of $Z_{n,m}(\mathfrak{R}[C])$, we get chain-maps: $e(\beta_1, \ldots, \beta_k) : Z_{n,m}(\mathfrak{R}[C]) \otimes \cdots \otimes Z_{n,m}(\mathfrak{R}[C]) \otimes C^k \rightarrow \mathcal{F}(C)^n$.

PROOF. This follows by direct computation — consider $\downarrow_{|\alpha|} \circ V(\beta_1, \ldots, \beta_k) \circ (\mathfrak{f}[\tilde{C}]_{\beta_1} \otimes \cdots \otimes \mathfrak{f}[\tilde{C}]_{\beta_k}) \circ V \circ (\uparrow^{|\beta_1|} \otimes \cdots \otimes \uparrow^{|\beta_k|}) \circ 1 \otimes \cdots \otimes 1 \otimes \Delta_k$. The sign of $(-1)^{k|\alpha|}$ (which cancels out the corresponding sign in the definition of $\bar{\mu}_k$) results from pushing the term $1 \otimes \cdots \otimes 1 \otimes \Delta_k$ past $\uparrow^{|\beta_1|} \otimes \cdots \otimes \uparrow^{|\beta_k|}$ and the conclusion follows from the definitions of t_k and m. $\square$

PROPOSITION 2.17. *For all* $n > 0$, $i > 1$ $(Z_{n,i}(\mathfrak{R}), \{\bar{\mu}_k, k \geq 2\})$, *equipped with the differential* $\hat{\mathfrak{T}}$, *constitutes an* $A(\infty)$-*algebra.*

REMARK. 2.17.1. We will ultimately form the bar construction of this $A(\infty)$-algebra and that will be the basis for giving a weakly coherent m-coalgebra structure to $\mathcal{F}(C)$.

PROOF. 1. Claim: When verifying the defining identity of an $A(\infty)$-algebra (namely 2.4 on page 40), the differential of $Z_{n,k}(\mathfrak{R})$ effectively behaves like $\bar{\mu}_1$. The terms of $\hat{\mathfrak{T}}$ contribute nothing — they cancel out due to the fact that they are defined in terms of left-composites of elements of $Z_{n,k}(\mathfrak{R})$ by the $\{\Delta_i\}$ and the $\{\bar{\mu}_j\}$ are defined in terms of right-composites. It turns out that the $\{\bar{\mu}_j\}$ essentially commute with the terms in $\hat{\mathfrak{T}}$ so that they cancel out in the defining identity of an $A(\infty)$-algebra. We verify this by direct computation — the typical higher term of $\hat{\mathfrak{T}} \circ \bar{\mu}_k$ that is applied to $r_1 \otimes \cdots \otimes r_k$ is:

$$(-1)^{i+|\alpha|j+ij} \downarrow^{|\alpha|+j-1} \circ (\Delta_j \circ_i *) \circ \uparrow^{|\alpha|} \downarrow^{\sum |\beta_j|} Z\{\beta_1, \ldots, \beta_k\} \circ t_k$$
$$\circ \uparrow^{|\beta_1|} \otimes \cdots \otimes \uparrow^{|\beta_k|} (r_1 \otimes \cdots \otimes r_k)$$

where $0 \le i \le |\alpha|$ and $2 \le j$ are some fixed integers and $\alpha = \sum_{i=1}^{k} \beta_i$ — this is equal to $(-1)^{i+|\alpha|j+ij} \downarrow^{|\alpha|+j-1} \circ (\Delta_j \circ_i *) \circ Z\{\beta_1, \ldots, \beta_k\} \circ \uparrow^{|\beta_1|}(r_1) \circ_1 \ldots \uparrow^{|\beta_k|}(r_k) \circ_k \Delta_k$, where $\hat{h}$ is equal to ± 1 (the sign that results from plugging $r_1 \otimes \cdots \otimes r_k$ into $\uparrow^{|\beta_1|} \otimes \cdots \otimes \uparrow^{|\beta_k|}$). This is equal to

$$\downarrow^{|\alpha|+j-1} \circ Z\{\beta_1, \ldots, \beta'_t, \ldots, \beta_k\} \circ (-1)^{i'+|\alpha|j+i'j} (\Delta_j \circ_z$$
$$\hat{h} \uparrow^{|\beta_1|}(r_1) \circ_1 \ldots \uparrow^{|\beta_t|}(r_t) \circ_t \ldots \uparrow^{|\beta_k|}(r_k) \circ_k \Delta_k)$$

where:

1. i' is equal to the largest value of k such that $i \ge \sum_{i=1}^{n}(\beta_i)_k$;
2. t is equal to the smallest value of k such that $i+1 - \sum_{i=1}^{n}(\beta_i)_{i'} \le \sum_{i=1}^{k}(\beta_i)_{i'}$.
3. $z = i' + \sum_{i=1}^{t-1}(\beta_i)_{i'}$;

The notation $(\beta_i)_{i'}$ denotes the i'^{th} term of the *sequence* β_i.

The sequence β'_t is the result of adding $j - 1$ to the i'^{th} term of the sequence β_t. The fact that we can permute $(-1)^{i+|\alpha|j+ij}\Delta_j \circ_i$ with $Z\{\beta_1, \ldots, \beta_k\}$ (with the signs working out properly) follows from the fact that we can regard $(-1)^{i+|\alpha|j+ij}\Delta_j \circ_i$ as $(-1)^{|\alpha|j+j}\Delta_j \circ_1$ as $(-1)^{|\alpha|j+j}\Delta_j \circ_1$, conjugated by the transposition that interchanges 1 and i. We now pass $\Delta_j \circ_z$ through the other terms to get

$$(-1)^{|\alpha|+j-1} \circ Z\{\beta_1, \ldots, \beta'_t, \ldots, \beta_k\} \circ (-1)^{i'+|\alpha|j+i'j}$$
$$\hat{h} \uparrow^{|\beta_1|}(r_1) \circ_1 \ldots (-1)^{j\sum_{u=1}^{t-1} \beta_u}\Delta_j \circ_{i'} \uparrow^{|\beta_t|}(r_t) \circ_t \ldots \uparrow^{|\beta_k|}(r_k) \circ_k \Delta_k)$$

$$= (-1)^{|\alpha|+j-1} \circ Z\{\beta_1, \ldots, \beta'_t, \ldots, \beta_k\} \circ (-1)^{i'+|\alpha|j+i'j}$$
$$\hat{h} \uparrow^{|\beta_1|}(r_1) \circ_1 \ldots (-1)^{j\sum_{u=1}^{t-1} \beta_u}\Delta_j \circ_{i'} \uparrow^{|\beta_t|}(r_t) \circ_t \ldots \uparrow^{|\beta_k|}(r_k) \circ_k \Delta_k)$$

But, except for a factor of $(-1)^{k-1}$ in the definition of t_k, this is the same as what we would have gotten by evaluating $(-1)^{j|\beta_t|+i'+i'j} \circ (\Delta_j \circ_{i'} *) \circ$ on $\uparrow^{|\beta_t|}(r_t)$ before plugging the result into $\bar{\mu}_k$. Basically, passing this term through $\uparrow^{|\beta_1|} \otimes \cdots \otimes \uparrow^{|\beta_k|}$ contributes a factor of $(-1)^{j\sum_{u=t+1}^{k} \beta_u}$ (i.e. it modifies $\hat{h}$) and that multiplies the sign of $(-1)^{|\beta_t|+i'+i'j}$ to give the same sign as appears above. This proves the initial claim

— when we form the expression $\sum_{k=1}^{n}\sum_{\lambda=1}^{n-k}(-1)^{k+\lambda+k\lambda}\bar{\mu}_{n-k+1}(1_\lambda\otimes\bar{\mu}_k\otimes 1_{n-k-\lambda})$ used in 2.4 on page 40, all contributions of $\widehat{\mathfrak{T}}$ will cancel out.

The proof that the remaining terms satisfy the properties of an $A(\infty)$-algebra follow immediately by duality and 2.16 on page 47 — the $\{\Delta_i\}$ already satisfy the corresponding identities for an $A(\infty)$-coalgebra. $\square$

DEFINITION 2.18. Let $\bar{\mathcal{B}}(Z_{n,m}(\mathfrak{R}))$ denote the formal bar construction — i.e the direct sum of copies of tensor-powers of $Z_{n,m}(\mathfrak{R}))$ with its associated product and coproduct-structures. This bar construction is formed with respect to the differential defined in 2.8 on page 42 (and proved to be a differential in 2.12 on page 43), and the $A(\infty)$-algebra structure $\{\bar{\mu}_k\}$, defined in 2.15 on page 45. The action of $\mathbb{Z}S_n$ on $Z_{n,m}(\mathfrak{R}))$ defined in remark 2.7.2 on page 42 is extended to $\bar{\mathcal{B}}(Z_{n,m}(\mathfrak{R})) - 2.15$ implies that this is well-defined.

In order to proceed any further we will need the following somewhat technical result:

LEMMA 2.19. Let $(f, g, \varphi)\colon (C, \partial_C) \to (D, \partial_D)$ be a contraction of chain-complexes and suppose C is equipped with a second differential ∂' such that the following conditions on the difference $t = \partial' - \partial_C$, are satisfied:

1. C is equipped with a filtration $F_i C$, such that $t(F_0 C) = 0$;
2. t and $\varphi \circ t$ lower filtration degree and $g \circ f$ preserves it;

Then $(f', g', \varphi')\colon (C, \partial') \to (D, \partial'')$ is a contraction of chain-complexes, where:

1. $f' = f \circ (1 + t \circ T_\infty \circ \varphi)$;
2. $g' = T_\infty \circ g$;
3. $\varphi' = T_\infty \circ \varphi$;
4. $\partial'' = \partial_D + f \circ t \circ T_\infty \circ g$; where

$$T_\infty = 1 + \sum_{i=1}^{\infty}(\varphi \circ t)^i$$

Remarks. 2.19.1. This result (or others equivalent to it) have appeared in a number of places over the last 20 years — see [4], [28], [10], and, more recently, [25], for instance. Our statement of the hypotheses is slightly different from the standard statement — we do not require that φ preserve filtration degree, only that $\varphi \circ t$ lower it. An examination of the proof of the perturbation lemma in [10] shows that our present hypotheses are sufficient.

2.19.2. Although T_∞ is written as an "infinite series", it reduces to a finite sum when evaluated on any element of C, due to the hypotheses.

PROPOSITION 2.20. For all $n, m > 1$ the bar constructions $\bar{\mathcal{B}}(Z_{n,m}(\mathfrak{R}))$, are acyclic in positive dimensions. In fact, as $\mathbb{Z}$-chain complexes, there exist contracting homotopies $\bar{\mathcal{B}}(\Phi_m)\colon \bar{\mathcal{B}}(Z_{n,m}(\mathfrak{R})) \to \bar{\mathcal{B}}(Z_{n,m}(\mathfrak{R}))$ that are compatible with the induced maps $\bar{\mathcal{B}}(p_{m',m})\colon \bar{\mathcal{B}}(Z_{n,m'}(\mathfrak{R})) \to \bar{\mathcal{B}}(Z_{n,m}(\mathfrak{R}))$ for all pairs m, m' where $m' > m$ (see 2.7 on

page 42)

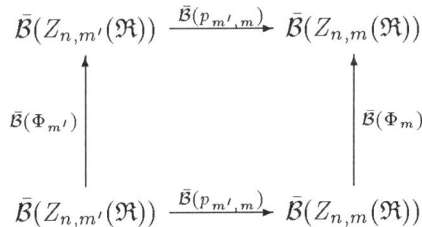

PROOF. If we only wanted to show that the chain-complexes were acyclic, we could simply use a spectral sequence argument. Since we also want the existence of compatible contracting cochains, we must work a little harder. Choose $\mathbb{Z}$-contracting cochains for all of the components of the f-resolution $\mathfrak{R}$:

$$\phi_n \colon \mathfrak{R}_n \to \mathfrak{R}_n$$

Since we are forgetting the group-actions on the chain-complexes, we can assume that there are actual *contractions* (see page 24) of chain-complexes

$$(\phi_n, i_n, \epsilon_n) \colon \mathfrak{R}_n \to V$$

where V is a chain-complex concentrated in dimension zero with $H_0(V) = \mathbb{Z}$. We get compatible systems of contractions of chain-complexes

$$(\Phi_{n,m'}, r_{n,m'}, x_{n,m'}) \colon Z_{n,m'}(\mathfrak{R}) \to V'$$

where "compatible" means that the maps $\Phi_{n,m'}$ commute with the maps $p_{m',m}$, and V' is concentrated in dimension -1. These contractions induce compatible contractions of tensor-algebras:

$$(T(\Sigma\Phi_{n,m'}), T(r_{n,m'}), T(x_{n,m'})) \colon T(\Sigma Z_{n,m'}) \to T(\Sigma V')$$

Now we regard these tensor-algebras as the "unperturbed" chain-complexes and regard the bar constructions as the *perturbed* chain-complexes. We can use the perturbation lemma above to prove the result if we can find a filtration of $\bar{\mathcal{B}}(Z_{n,m}(\mathfrak{R}))$ in such a way that the boundary maps and the contracting cochains preserve filtration degree. We define the filtration-degree of a canonical basis element $x = [a_1 | \dots | a_k]$, with $a_i \in \mathfrak{R}_{z_i}$, to be $\sum_{i=1}^k z_i$. $\square$

PROPOSITION 2.21. *Consider the inverse system* $p_{m',m} \colon Z_{n,m'}(\mathfrak{R}) \to Z_{n,m}(\mathfrak{R})$. *This gives rise to an inverse system* $\bar{\mathcal{B}}(Z_{n,m'}(\mathfrak{R})) \to \bar{\mathcal{B}}(Z_{n,m}(\mathfrak{R}))$. *The inverse limits,* $\bar{\mathcal{B}}(\widehat{Z_{n,*}(\mathfrak{R}}))= \varprojlim \bar{\mathcal{B}}(Z_{n,m}(\mathfrak{R}))$, *are acyclic in positive dimensions.*

PROOF. We must show that the derived functor of the inverse limit vanishes in this case.

Suppose that $k > 0$ and

$$\cdots \mapsto a_m \mapsto \cdots$$

is a cycle in $\varprojlim \bar{\mathcal{B}}(Z_{n,m}(\mathfrak{R}))$, where $a_m \in \bar{\mathcal{B}}(Z_{n,m}(\mathfrak{R}))_k$ and $p_{m',m}a_{m'} = a_m$ for all m, m'.

Each of the a_m will be cycles in their respective $\bar{\mathcal{B}}(Z_{n,m}(\mathfrak{R}))$. We must find elements $c_m \in \bar{\mathcal{B}}(Z_{n,m}(\mathfrak{R}))_{k+1}$ such that

(2.2) $\partial(c_m) = a_m$

(2.3) $p_{m',m}(c_{m'}) = c_m$ for all pairs m, m' such that $m' > m$

The conclusion follows from 2.20 on page 49, which implies that we can choose $c_m = \bar{\mathcal{B}}(\Phi_m)(a_m)$ for all $m > 1$.   □

PROPOSITION 2.22. *The completion* $\widehat{\bar{\mathcal{B}}(Z_{n,*}(\mathfrak{R}))}$ *comes equipped with a coproduct* $\hat{\Delta}\colon \widehat{\bar{\mathcal{B}}(Z_{n,*}(\mathfrak{R}))} \to \widehat{\bar{\mathcal{B}}(Z_{n,*}(\mathfrak{R}))} \otimes \widehat{\bar{\mathcal{B}}(Z_{n,*}(\mathfrak{R}))}$ *that induces the standard coproducts on the* $\bar{\mathcal{B}}(Z_{n,m}(\mathfrak{R}))$ *for all* $m > 1$.

REMARK. 2.22.1. Recall the standard coproduct on $\bar{\mathcal{B}}(Z_{n,m}(\mathfrak{R}))$: it is defined by the following formula on canonical basis elements:

$$\Delta_{\bar{\mathcal{B}}(Z_{n,m}(\mathfrak{R}))}([a_1|\ldots|a_k]) = \sum_{i=0}^{k} [a_1|\ldots|a_i] \otimes [a_{i+1}|\ldots|a_k]$$

where $[a_1|\ldots|a_0] = [\,]$ and $[a_{m+1}|\ldots|a_m] = [\,]$. This is extended $\mathbb{Z}$-linearly to all of $\bar{\mathcal{B}}(Z_{n,*}(\mathfrak{R}))$.

PROOF. This follows from the easily-verified fact that each of the standard coproducts commutes with the maps $\{p_{m',m}\}$:

$$\begin{array}{ccc}
\bar{\mathcal{B}}(Z_{n,m'}(\mathfrak{R})) & \xrightarrow{\;\bar{\mathcal{B}}(p_{m',m})\;} & \bar{\mathcal{B}}(Z_{n,m}(\mathfrak{R})) \\
\Big\downarrow{\scriptstyle \Delta_{\bar{\mathcal{B}}(Z_{n,m'}(\mathfrak{R}))}} & & \Big\downarrow{\scriptstyle \Delta_{\bar{\mathcal{B}}(Z_{n,m}(\mathfrak{R}))}} \\
\bar{\mathcal{B}}(Z_{n,m'}(\mathfrak{R})) \otimes \bar{\mathcal{B}}(Z_{n,m'}(\mathfrak{R})) & \xrightarrow[\bar{\mathcal{B}}(p_{m',m}) \otimes \bar{\mathcal{B}}(p_{m',m})]{} & \bar{\mathcal{B}}(Z_{n,m}(\mathfrak{R})) \otimes \bar{\mathcal{B}}(Z_{n,m}(\mathfrak{R}))
\end{array}$$

It follows that we can define the coproduct of

$$\{\cdots \mapsto a_m \mapsto \cdots \mapsto a_1\}$$

to be

$$\{\cdots \mapsto \Delta_{\bar{\mathcal{B}}(Z_{n,m}(\mathfrak{R}))}(a_m) \mapsto \cdots \mapsto \Delta_{\bar{\mathcal{B}}(Z_{n,1}(\mathfrak{R}))}(a_1)\} \subset \widehat{\bar{\mathcal{B}}(Z_{n,*}(\mathfrak{R}))} \otimes \widehat{\bar{\mathcal{B}}(Z_{n,*}(\mathfrak{R}))}$$

□

Now we are in a position to define the coordinate maps of the cobar construction.

DEFINITION 2.23. Consider   the   canonical   isomorphism
$s \colon \mathrm{Hom}_{\mathbb{Z}}(\mathfrak{R}[C], C) \quad \to \quad \mathrm{Hom}_{\mathbb{Z}}(\Sigma^{-1}\mathfrak{R}[C], \Sigma^{-1}C)$ that sends a map
$f \in \mathrm{Hom}_{\mathbb{Z}}(\mathfrak{R}[C], C)$ to $\downarrow \circ f \circ \uparrow \in \mathrm{Hom}_{\mathbb{Z}}(\Sigma^{-1}\mathfrak{R}[C], \Sigma^{-1}C)$. Define $s(n)$ to be the
canonical isomorphism $s(n) \colon \mathrm{Hom}_{\mathbb{Z}}(\mathfrak{R}[C], C) \quad \to \quad \mathrm{Hom}_{\mathbb{Z}}(\Sigma^{-n}\mathfrak{R}[C], (\Sigma^{-1}C)^n)$
— where we identify $(\Sigma^{-1}C)^n$ with $\Sigma^{-1}C^n$. If $g \in \mathrm{Hom}_{\mathbb{Z}}(\mathfrak{R}[C], C)$, then
$s(n)(g) = \downarrow_n \circ g \circ \uparrow^n$. It is essentially s^n composed with a suitable shuffle-map. The
notation $\downarrow_n$ is defined in 2.10 on page 43.

DEFINITION 2.24. Suppose $\quad (C, \quad \{\mathfrak{f}[C]_n \colon C \quad \to \quad \mathrm{Hom}_{\mathbb{Z}S_n}(\mathfrak{R}[C]_n, C^n)\})$
is a weakly-coherent m-coalgebra. For all $n \geq 0$, define the maps
$\mathcal{F}(\mathfrak{f}[C]_n)' \colon \mathcal{F}(C) \to \mathrm{Hom}_{\mathbb{Z}S_n}(\bar{\mathcal{B}}(\widehat{Z_{n,*}(\mathfrak{R}[C])}), \mathcal{F}(C)^n)$ by:

1. $\mathcal{F}(\mathfrak{f}[C]_n)' \quad = \quad \sum_\alpha s(|\alpha|) \quad \circ \quad (\mathfrak{f}[C]_{|\alpha|} \circ \downarrow) \circ \uparrow \colon \Sigma^{-1}C \quad \to$
   $\mathrm{Hom}_{\mathbb{Z}S_n}(\bar{\mathcal{B}}(Z_{n,m}(\mathfrak{R}[C])), \mathcal{F}(C)^n)$. Here the target of the term
   $s(|\alpha|) \circ (\mathfrak{f}[C]_{|\alpha|} \circ \downarrow) \circ \uparrow$ lies in $(\Sigma^{-1}C)^{\alpha_1} \otimes \cdots \otimes (\Sigma^{-1}C)^{\alpha_n} \subset \mathcal{F}(C)^n$,
   and the summation is taken over all length-n sequences, α, of nonnegative
   integers. The definition of the action of S_n on $Z_{n,m}(\mathfrak{R}[C])$ given in remark
   2.7.2 on page 42 implies that the desuspension of $\mathfrak{f}[C]_{|\alpha|}$ is still $\mathbb{Z}S_n$-linear.

2. Requiring that $\mathcal{F}(C)$ be an m-Hopf algebra, as defined in 3.13 on page 26.

Remarks. 2.24.1. The expression $\sum_\alpha s(|\alpha|) \circ (\mathfrak{f}[C]_{|\alpha|} \circ \downarrow) \circ \uparrow$ is formally an in-
finite sum. It is well-defined and has its target in a finite-dimensional submodule of
$\bar{\mathcal{B}}(Z_{n,m}(\mathfrak{R}[C]))$ (actually $\Sigma Z_{n,m}(\mathfrak{R}[C])$) because of condition 1c in 3.3 on page 19.

2.24.2. $s(|\alpha|) \circ (\mathfrak{f}[C]_{|\alpha|} \circ \downarrow) \circ \uparrow$ is interpreted as follows: the $\uparrow$-factor is a
map $\uparrow \colon \Sigma^{-1}C \to C$, and the $\downarrow$-factor sends $\Sigma Z_{n,m}(\mathfrak{R}[C]) \subset \bar{\mathcal{B}}(Z_{n,m}(\mathfrak{R}[C]))$ to
$Z_{n,m}(\mathfrak{R}[C])$.

2.24.3. Recall the the condition that $\mathcal{F}(C)$ forms an m-Hopf algebra implies that
$\mathfrak{f}[C]_n(x_1 \cdot x_2)(\delta) = \cdot f_i(x_1) \otimes f_j(x_2)(\Delta_R(\delta))$, where $\cdot \colon \mathcal{F}(C) \otimes \mathcal{F}(C) \to \mathcal{F}(C)$ is
the (tensor) product defined on the cobar construction, and $\Delta_R \colon \bar{\mathcal{B}}(\widehat{Z_{n,*}(\mathfrak{R}[C])}) \to$
$\bar{\mathcal{B}}(\widehat{Z_{n,*}(\mathfrak{R}[C])}) \otimes \bar{\mathcal{B}}(\widehat{Z_{n,*}(\mathfrak{R}[C])})$ is the coproduct defined in 2.22 on page 51 (and x_1,
x_2 are elements of $\mathcal{F}(C)$ and δ is an element of $\mathfrak{R}[C]_n$). By abuse of notation, we
regard the tensor-multiplication as being defined over $(\mathcal{F}(C))^n$ as well as $\mathcal{F}(C)$ — we
use the usual Koszul convention in this case.

Such a definition makes sense because:

1. $\mathcal{F}(C)$ is a free DGA algebra;
2. the product $\cdot$ is associative;
3. the coproduct Δ_R is co-associative.

This essentially implies that $f_i(x_1 \otimes \cdots \otimes x_k)$ has its values in maps whose support
is in $\Sigma Z_{i,t}(\mathfrak{R}[C]) \otimes \cdots \otimes \Sigma Z_{i,t}(\mathfrak{R}[C])$ (k factors) $\subset \bar{\mathcal{B}}(Z_{n,t}(\mathfrak{R}[C]))$ (technically, it
means that the maps have their image in portions of $\bar{\mathcal{B}}(\widehat{Z_{n,*}(\mathfrak{R}[C])})$ that factor through
$\Sigma Z_{i,t}(\mathfrak{R}[C]) \otimes \cdots \otimes \Sigma Z_{i,t}(\mathfrak{R}[C])$, for suitable values of t).

THEOREM 2.25. *The maps* $\mathcal{F}(\mathfrak{f}[C]_n)' \colon \mathcal{F}(C) \to \mathrm{Hom}_{\mathbb{Z}S_n}(\bar{\mathcal{B}}(\widehat{Z_{n,*}(\mathfrak{R}[C])}), \mathcal{F}(C)^n)$,
defined in definition 2.24 on page 52, are chain-maps for all $n \geq 1$.

PROOF. Without loss of generality, we replace $\bar{B}(Z_{n,*}\widehat{(\mathfrak{R}[C])})$ by $\bar{B}(Z_{n,t}(\mathfrak{R}[C]))$ for a suitable value of t. We prove the statement of this result in three steps:

1. First, suppose that $\mathcal{F}(C) = T(\Sigma^{-1}C)$ — the tensor algebra of the de-suspension of C; $Z_{n,t}(\mathfrak{R}[C]) =$ the result of truncating $\Sigma^{-|\alpha|}\mathfrak{R}[C]^{|\alpha|}$ in dimension -1; and $\bar{B}(Z_{n,t}(\mathfrak{R}[C])) = T(\Sigma Z_{n,t}(\mathfrak{R}[C]))$ — the tensor-algebra of the suspension of $Z_{n,t}(\mathfrak{R}[C])$. Then the maps

$$\mathcal{F}(\mathfrak{f}[C]_n)' : T(\Sigma^{-1}C) \to \mathrm{Hom}_{\mathbb{Z}S_n}(T(\Sigma Z_{n,t}(\mathfrak{R}[C])), T(\Sigma^{-1}C)^n)$$

are clearly chain-maps — this is an immediate consequence of the fact that the maps $\{\mathfrak{f}[C]_n : C \to \mathrm{Hom}_{\mathbb{Z}S_n}(\mathfrak{R}[C]_n, C^n)\}$ are chain-maps. The actual differentials in the objects under consideration (namely $\bar{B}(Z_{n,t}(\mathfrak{R}[C]))$, $Z_{n,t}(\mathfrak{R}[C])$, and $\mathcal{F}(C)$) are perturbations of these differentials, and we have to verify that these perturbations are compatible. This follows from:

2. Consider the situation:

$$\mathcal{F}(\mathfrak{f}[C]_n)' : T(\Sigma^{-1}C) \to \mathrm{Hom}_{\mathbb{Z}S_n}(T(\Sigma Z_{n,t}(\mathfrak{R}[C])), \mathcal{F}(C)^n)$$

where $Z_{n,t}(\mathfrak{R}[C])$ now has the differential defined in 2.8 on page 42.

CLAIM 2.25.1. *In this case, the maps $\mathcal{F}(\mathfrak{f}[C]_n)'$ are chain-maps.*

Proof of claim: We prove that $\mathcal{F}(\mathfrak{f}[C]_n)'|\Sigma^{-1}C : \Sigma^{-1}C \to \mathrm{Hom}_{\mathbb{Z}S_n}(\Sigma Z_{n,t}(\mathfrak{R}[C]), \mathcal{F}(C)^n)$ is a chain-map. We will now compute the boundary of an element $z \in \mathrm{Hom}_{\mathbb{Z}S_n}(\Sigma Z_{n,t}(\mathfrak{R}[C]), \mathcal{F}(C)^n)$ and show that it is the same as the boundary of z, regarded as an element of $\mathrm{Hom}_{\mathbb{Z}S_n}(T(\Sigma Z_{n,t}(\mathfrak{R}[C])), T(\Sigma^{-1}C)^n)$ in case 1 above. The conclusion will follow from case 1 above. Suppose $\{\Delta_i : C \to C^i\}$ stands for the canonical $A(\infty)$-coalgebra structure induced by its weakly-coherent m-structure — see 1.5 on page 36. The boundary of z is $\partial_{\mathcal{F}} \circ z - (-1)^{\dim(z)} z \circ \partial$. The composition

$$\underbrace{1 \otimes \cdots \otimes (\downarrow_i \circ \Delta_i \circ \uparrow) \otimes \cdots \otimes 1}_{j^{\text{th}} \text{ position}} \circ s(|\alpha|) \circ \mathfrak{f}[C]_{|\alpha|}$$

$$= \underbrace{1 \otimes \cdots \otimes (\downarrow_i \circ \Delta_i \circ \uparrow) \otimes \cdots \otimes 1}_{j^{\text{th}} \text{ position}} \circ \downarrow_{|\alpha|} \circ \mathfrak{f}[C]_{|\alpha|} \circ \uparrow^{|\alpha|}$$

Recall the notation $\downarrow_i = \underbrace{\downarrow \otimes \ldots \downarrow}_{i \text{ copies}}$ — see 2.10 on page 43. We slide $\downarrow_i \circ \Delta_i \circ \uparrow$ past $j - 1$ copies of $\downarrow$ to get a sign of $(-1)^{j-1}$ (since $\downarrow$ and $\downarrow_i \circ \Delta_i \circ \uparrow$ are both of degree -1). The factor of $\uparrow$ cancels a corresponding factor of $\downarrow$. Now we slide the Δ_i past the remaining $|\alpha| - j$ copies of $\downarrow$ to get another factor of $(-1)^{i(|\alpha|-j)}$. This is due to the fact that Δ_i is of degree $i - 2 \equiv i \pmod{2}$, and $\downarrow$ is of degree -1. To summarize so far, we have:

$$\underbrace{1 \otimes \cdots \otimes (\downarrow_i \circ \Delta_i) \circ \uparrow \otimes \cdots \otimes 1}_{j^{\text{th}} \text{ position}} \circ s(|\alpha|) \circ f[C]_{|\alpha|}$$

$$= -(-1)^{j+|\alpha|i+ij} \downarrow_{|\alpha|+i-1} \circ \underbrace{1 \otimes \cdots \otimes \Delta_i \otimes \cdots \otimes 1}_{j^{\text{th}} \text{ position}} \circ f[C]_{|\alpha|} \circ \uparrow^{|\alpha|}$$

Recall lemma 1.6 on page 37: we have the following identity when evaluating the structure maps on an element, $c \in C$:

$$\underbrace{1 \otimes \cdots \otimes \Delta_i \otimes \cdots \otimes 1}_{j^{\text{th}} \text{ position}} \circ f[C]_{|\alpha|}(c) = (-1)^{\dim(c)i} f[C]_{|\alpha|+i-1} \circ (\Delta_i \circ_j *)$$

The boundary of an element of $z \subset \operatorname{Hom}_{\mathbb{Z}S_n}(T(\Sigma Z_{n,t}(\mathfrak{R}[C])), \mathcal{F}(C)^n)$ is induced by its boundary in $\operatorname{Hom}_{\mathbb{Z}}(T(\Sigma Z_{n,y}(\mathfrak{R}[C])), \mathcal{F}(C)^n)$, which is $\partial_{\mathcal{F}} \circ z - (-1)^{\dim(z)} z \circ \partial Z$. In the present situation

(2.4)
$$\partial_{\mathcal{F}} \circ z(c) = -(-1)^{j+|\alpha|i+ij} \downarrow_{|\alpha|+i-1}$$

$$\circ \underbrace{1 \otimes \cdots \otimes \Delta_i \otimes \cdots \otimes 1}_{j^{\text{th}} \text{ position}} \circ f[C]_{|\alpha|} \circ \uparrow^{|\alpha|}(c)$$

$$= -(-1)^{j+|\alpha|i+ij}(-1)^{\dim(c)i}(-1)^{\dim(c)|\alpha|} \downarrow_{|\alpha|+i-1} \circ f[C]_{|\alpha|} \circ (\Delta_i \circ_j *) \circ \uparrow^{|\alpha|}$$

(the extra factor of $(-1)^{\dim(c)|\alpha|}$ resulted from slipping c past the term of $\uparrow^{|\alpha|}$); and

(2.5) $$(-1)^{\dim(c)} z \circ \partial_Z s(|\alpha|) \circ f[C]_{|\alpha|}(c)$$

$$= -(-1)^{j+|\alpha|i+ij} s(|\alpha|) \circ f[C]_{|\alpha|}(c) \downarrow^{|\alpha|+i-1} \circ (\Delta_i \circ_j *) \circ \uparrow^{|\alpha|}$$

$$= -(-1)^{\dim(c)(|\alpha|+i)}(-1)^{j+|\alpha|i+ij} \downarrow^{|\alpha|+i-1} s(|\alpha|) \circ f[C]_{|\alpha|}(c) \circ (\Delta_i \circ_j *) \circ \uparrow^{|\alpha|}$$

since we have slipped $\downarrow^{|\alpha|+i-1}$ past $s(|\alpha|) \circ f[C]_{|\alpha|}(c)$. The difference of these two terms, (2.4) and (2.5), is clearly zero.

3. Now we consider the full statement of the result. We have to verify that the perturbation of the boundary of $T(\Sigma^{-1}C)$ that results when it is replaced by $\mathcal{F}(C)$ is compatible with the perturbation of the boundary of $\operatorname{Hom}_{\mathbb{Z}S_n}(T(\Sigma Z_{n,t}(\mathfrak{R}[C])), \mathcal{F}(C)^n)$ when it is replaced by $\operatorname{Hom}_{\mathbb{Z}S_n}(\bar{\mathcal{B}}(Z_{n,t}(\mathfrak{R}[C])), \mathcal{F}(C)^n)$. This follows from:

   a. the fact that the perturbation of the boundary of $\bar{\mathcal{B}}(Z_{n,t}(\mathfrak{R}[C]))$ is simply added to the boundary of an element of $\operatorname{Hom}_{\mathbb{Z}S_n}(T(\Sigma Z_{n,t}(\mathfrak{R}[C])), \mathcal{F}(C)^n)$; and

   b. 2.16 on page 47.

This completes the proof. $\square$

We are now in a position to prove the main computational results of this section:

THEOREM 2.26. *For all n there exist homomorphisms of coalgebras* $h_n \colon R(S_n) \to \widehat{\bar{\mathcal{B}}(Z_{n,*}(\mathfrak{R}))}$. *These homomorphisms are computed as follows:*

1. *compute twisting cochains* $\sum_{|\alpha| \le k} z_\alpha \colon R(S_n) \to Z_{n,k}(\mathfrak{R})$, *where* $z_\alpha \colon R(S_n) \to \mathfrak{R}_{|\alpha|}$ *are suitable functions, and compute homomorphisms of coalgebras* $b(\sum_\alpha \downarrow^{|\alpha|} \circ z_\alpha) \colon R(S_n) \to \bar{\mathcal{B}}(Z_{n,k}(\mathfrak{R}))$ *— these homomorphisms are compatible with the maps in the inverse system* $\{\bar{\mathcal{B}}(p_{k',k}) \colon \bar{\mathcal{B}}(Z_{n,k'}(\mathfrak{R})) \to \bar{\mathcal{B}}(Z_{n,k}(\mathfrak{R}))\}$;

2. *for all n this defines homomorphisms of coalgebras* $b(\sum_\alpha \downarrow^{|\alpha|} \circ z_\alpha) \colon R(S_n) \to \varprojlim \bar{\mathcal{B}}(Z_{n,k}(\mathfrak{R}))$;

REMARK. 2.26.1. It turns out that there doesn't exist a twisting cochain $z \colon R(S_n) \to Z_{n,\infty}(\mathfrak{R})$ — we can only define *compatible* maps to *quotients* $Z_{n,k}(\mathfrak{R})$. This is the basic reason we work with the inverse limits, $\varprojlim \bar{\mathcal{B}}(Z_{n,k}(\mathfrak{R}))$.

PROPOSITION 2.27. *Let* $z_n(R(S_n)) = \sum_\alpha \downarrow^{|\alpha|} \circ z_\alpha$, *where the summation is taken over all sequences of nonnegative integers of length n, and* $z_\alpha \colon R(S_n) \to \mathfrak{R}_{|\alpha|}$ *are maps of degree* $|\alpha| - 1$. *Then, on canonical basis elements of $R(S_n)$ the z_α must satisfy the equation:*

$$z_\alpha \circ \partial_R + (-1)^{|\alpha|} \partial_{\mathfrak{R}} \circ z_\alpha$$
$$+ \sum_{i=2}^\infty (-1)^{i|\alpha|} \sum_{\beta_1 + \cdots + \beta_i = \alpha} (-1)^{\sum_{j=2}^i \left(|\beta_j| \sum_{t=1}^{j-1} (|\beta_t| - 1) \right)}$$
$$\cdot \mathcal{Z}\{\beta_1, \ldots, \beta_i\} \circ t_i \circ z_{\beta_1} \otimes \cdots \otimes z_{\beta_i} \circ \Delta_{\mathfrak{R}}^{i-1}$$
$$+ \sum_{i=1}^{|\alpha|} \sum_{j=2}^\infty (-1)^{i+|\alpha|j+ij} \Delta_j \circ_i z_\alpha(i,j)$$
$$= 0$$

Remarks. 2.27.1. Here the $\{\beta_i\}$ are lists of length n of nonnegative integers and the sum $\beta_1 + \cdots + \beta_i = \alpha$, in the second line of the formula, is taken elementwise.

2.27.2. This is a straightforward consequence of definition 2.5 on page 40; the definition of the $\{\bar\mu_k\}$ in the remarks following definition 2.14 on page 45; the definition of $\widehat{\mathfrak{T}}$ in 2.8 on page 42; and the Koszul convention. Let $\gamma(i)$ be defined to be the largest value of t such that $\sum_{k=1}^{t-1} \alpha_k < i \le \sum_{k=1}^t \alpha_k$.

Then $z_\alpha(i,j)$ is defined to be $\begin{cases} 0 & \text{if } \alpha_{\gamma(i+j-1)} \le j-1 \\ z_{\alpha'} & \text{if } \alpha_{\gamma(i+j-1)} > j-1 \end{cases}$, where α' is the sequence that results from subtracting $j-1$ from the $\gamma(i+j-1)^{\text{th}}$ term of α.

2.27.3. We can use this formula to inductively compute the $\{z_\alpha\}$. If we order the sequences of length n lexicographically, this formula expresses boundaries of z_α's in terms of $z_{\alpha'}$'s where the α' are strictly less than α in this ordering.

2.27.4. The $\{t_i\}$ are defined in remark 2.14.2 on page 45.

This inductive computation is begun by requiring that $z_\alpha \colon R(S_1) \to \Sigma^{-1}\mathfrak{R}_\alpha = Z_{1,1}(\mathfrak{R})$, sends $1 \in R(S_1) = \mathbb{Z}$ to $\Sigma^{-1}1$, whenever α is a sequence whose single nonzero entry is equal to 1. This procedure defines the $\{z_\alpha\}$ on canonical basis elements of $R(S_n)$ — the definition is extended to all of $R(S_n)$ by defining $z_{\{\alpha_1,\dots,\alpha_n\}}(\sigma \cdot a) = \text{Parity}((\sigma))(\sigma) \cdot z_{\sigma^{-1}\{\alpha_1,\dots,\alpha_n\}}(a)$, where a is a canonical basis element and $\sigma \in S_n$, and $\text{Parity}((\sigma)) = \{\pm 1\}$. The notation $\sigma^{-1}\{\alpha_1,\dots,\alpha_n\}$ means "permute the list of n elements $\{\alpha_1,\dots,\alpha_n\}$ via σ^{-1}", i.e. $\sigma^{-1}\{\alpha_1,\dots,\alpha_n\} = \{\alpha_{\sigma(1)},\dots,\alpha_{\sigma(n)}\}$. The map $T_{\alpha_1,\dots,\alpha_n} \colon S_n \to S_{|\alpha|}$ was defined in 2.7 on page 14 — its main property of interest to us is that $T_{\alpha_1,\dots,\alpha_n}(\sigma_1 \cdot \sigma_2) = T_{\alpha_1,\dots,\alpha_n}(\sigma_1) \cdot T_{\sigma_1^{-1}\{\alpha_1,\dots,\alpha_n\}}(\sigma_2)$ — see 2.9 on page 15. The definition of the action of $\mathbb{Z}S_n$ on $Z_{n,m}(\mathfrak{R})$ in remark 2.7.2 on page 42 implies that the sum of the $\{z_\alpha\}$ constitutes a $\mathbb{Z}S_n$-linear map $R(S_n) \to Z_{n,m}(\mathfrak{R})$.

PROPOSITION 2.28. *Let $(C, \{\mathfrak{f}[C]_n \colon C \to \text{Hom}_{\mathbb{Z}S_n}(\mathfrak{R}[C]_n, C^n)\})$ be a weakly coherent m-coalgebra, let $h_n \colon R(S_n) \to \bar{\mathcal{B}}(Z_{n,*}(\mathfrak{R}[C]))$ be the homomorphisms of coalgebras computed in 2.26 on page 55 and let $H_0(h_n) \colon \mathbb{Z} \to H_0(\bar{\mathcal{B}}(Z_{n,*}(\mathfrak{R}[C])))$ be the induced map in H_0. Since there exists a natural projection $\bar{\mathcal{B}}(Z_{n,*}(\mathfrak{R}[C])) \to H_0(\bar{\mathcal{B}}(Z_{n,*}(\mathfrak{R}[C])))$, we may form the fibered product with respect to these two maps to $H_0(\bar{\mathcal{B}}(Z_{n,*}(\mathfrak{R}[C])))$. Call this fibered product $\mathcal{F}(\mathfrak{R}[C], n)$. This:*

1. *is a $\mathbb{Z}S_n$-resolution of $\mathbb{Z}$;*
2. *is a DGA-sub-coalgebra of $\bar{\mathcal{B}}(Z_{n,*}(\mathfrak{R}[C]))$;*
3. *fits into a sequence of chain-maps $\mathcal{F}(\mathfrak{f}[C]_n) \colon \mathcal{F}(C) \to \text{Hom}_{\mathbb{Z}S_n}(\mathcal{F}(\mathfrak{R}[C], n), \mathcal{F}(C)^n)$, where $\mathcal{F}(\mathfrak{f}[C]_n) = r \circ \sum_\alpha s^{|\alpha|} \circ (\mathfrak{f}[C]_{|\alpha|} \circ \downarrow) \circ \uparrow$ and $r \colon \text{Hom}_{\mathbb{Z}S_n}(\bar{\mathcal{B}}(Z_{n,*}(\mathfrak{R}[C])), \mathcal{F}(C)^n) \to \text{Hom}_{\mathbb{Z}S_n}(\mathcal{F}(\mathfrak{R}[C], n), \mathcal{F}(C)^n)$ is induced by restriction.*

In addition, the maps $h_n \colon R(S_n) \to \bar{\mathcal{B}}(Z_{n,*}(\mathfrak{R}[C]))$ induce maps[1] $h_n \colon R(S_n) \to \mathcal{F}(\mathfrak{R}[C], n)$.

Remarks.   2.28.1. All of these statements follow from the basic properties of fibered products or have been proved already in definition 2.24 on page 52, 2.25 on page 52, 2.26 on page 55, or 2.21 on page 50.

2.28.2. the $\{\mathcal{F}(\mathfrak{R}[C], n)\}$ defined above will be the coordinate coalgebras of the cobar construction. They clearly come equipped with canonical maps $\mathfrak{R}[C]_n \to \mathcal{F}(\mathfrak{R}[C], n)$ that induce homology isomorphisms. All that still remains to be done is the definition of composition operations:

DEFINITION 2.29. Let $(C, \{\mathfrak{f}[C]_n \colon C \to \text{Hom}_{\mathbb{Z}S_n}(\mathfrak{R}[C]_n, C^n)\})$ be a weakly coherent m-coalgebra.

1. Let $\alpha = \{\alpha_1,\dots,\alpha_n\}$, $\beta = \{\beta_1,\dots,\beta_m\}$ be sequences of nonnegative integers. If $1 \leq i \leq m$ and $1 \leq j \leq \alpha_i$ we may define composition-operations

$$\circ_{i,j} = \downarrow^{|\gamma|} \circ_{i'} \uparrow^{|\alpha|} \otimes \uparrow^{|\beta|} \colon \Sigma^{-|\alpha|}\mathfrak{R}[C]_{|\alpha|} \otimes \Sigma^{-|\beta|}\mathfrak{R}[C]_{|\beta|}$$
$$\to \Sigma^{-|\gamma|}\mathfrak{R}[C]_{|\gamma|} \subset Z_{m+n-1}(\mathfrak{R}[C])$$

---

[1] Denoted by the same name, by abuse of notation.

where $i' = j + \sum_{k=1}^{i-1} \alpha_k$. The list $\gamma = \{\gamma_1, \ldots, \gamma_{m+n-1}\}$ is defined as follows:

Let $t(i)$ be largest value[2] such that $\sum_{j=1}^{t(i)} \beta_j \leq i < \sum_{j=1}^{t(i)+1} \beta_j$ — here we follow the convention that $\sum_{j=1}^{0} \beta_j = 0$. We consider several cases (where in all cases $\gamma_j = \beta_j$, $j < t(i)$):

    a. $\sum_{j=1}^{t(i)} \beta_j = i$. Then $\gamma_{t(i)} = \beta_{t(i)} + \alpha_1 - 1$, and $\gamma_{j+t(i)} = \alpha_{j+1}$, $1 \leq j \leq n-1$, and $\gamma_{j+t(i)+n} = \beta_{j+t(i)}$, $1 \leq j \leq m - t(i)$;

    b. $\sum_{j=1}^{t(i)} \beta_j < i$. Then $\gamma_{t(i)} = \beta_{t(i)}$; $\gamma_{t(i)+1} = i - \sum_{j=1}^{t(i)} \beta_j + \alpha_1 - 2$; and

$$\gamma_{j+t(i)+1} = \alpha_{j+1}, 1 \leq j \leq n-2, \gamma_{t(i)+n} = \alpha_n + \sum_{j=1}^{t(i)+1} \beta_j - i - 1$$

, and $\gamma_{j+t(i)+n} = \beta_{j+t(i)}$, $2 \leq j \leq m - t(i)$.

2. It is possible to define a formal coalgebra structure $\circ_i \colon \mathcal{F}(\mathfrak{R}[C], m) \otimes \mathcal{F}(\mathfrak{R}[C], n) \to \mathcal{F}(\mathfrak{R}[C], n + m - 1)$, on $\{\mathcal{F}(\mathfrak{R}[C], n)\}$ as follows: Let $a = [r_1 | \ldots | r_k] \in \mathcal{F}(\mathfrak{R}[C], n) \subset \bar{B}(Z_{n,t}(\mathfrak{R}[C]))$ be an element with $r_j \in \Sigma^{-|\alpha_j|} \mathfrak{R}[C]_{\alpha_j}$. Then $b \circ_i a$, where b is an element of $\bar{B}(Z_{m,t}(\mathfrak{R}[C]))$, that is nonzero if and only if b is of the form $[s_{1,1} | \ldots | s_{1,\alpha_{1,i}} | s_{2,1} | \ldots | s_{2,\alpha_{2,i}} | \ldots | s_{k,1} | \ldots | s_{k,\alpha_{k,i}}]$ — in which case the composite is equal to

$$(-1)^{(m-1)C} \cdot [r_1' | \ldots | r_k']$$

where $r_t' = s_{t,1} \circ_{i,1} r_t$, and $C = \sum_{j=1}^{i-1} \alpha_{1,j} + |\alpha_1| + \sum_{j=1}^{i-1} \alpha_{2,j} + |\alpha_1| + |\alpha_2| + \cdots + \sum_{j=1}^{k-1} |\alpha_i| + \sum_{j=1}^{i-1} \alpha_{k,j} = \sum_{t=1}^{k} \sum_{j=1}^{i-1} \alpha_{t,j} + \sum_{j=1}^{k-1} (k-j)|\alpha_j|$.

Remarks. 2.29.1. The notation $\alpha_{i,j}$ refers to the jth element of the list α_i. The condition on b is nontrivial — it implies, for instance, that the number of factors in b is equal to $\sum_{j=1}^{k} \alpha_{j,i}$.

2.29.2. Although the definition of the $\{\circ_{i,j}\}$ is fairly complex the resulting definition of the composition-operations on the $\{\mathcal{F}(\mathfrak{R}[C], n)\}$ is somewhat simpler. Basically, $a \circ_{i,j} b$ represents the following composition: regard the target of b (remember that the elements of $Z_{n,m}(\mathfrak{R}[C])$ are elements of a formal coalgebra, hence formal composites of elements that represent maps of C into tensor-powers of C) as a tensor-product of elements of $T(C)$ (tensor algebra on some chain-complex C). Now plug a into the jth copy of C in the ith factor of of this product of elements of $T(C)$.

The meaning of the description of γ is as follows: regard the term $\beta_{t(i)}$ in b as a sequence of $\beta_{t(i)}$ copies of 1 (each copy of 1 represents a copy of C in the $t(i)^{\text{th}}$ copy of $T(C)$, as per the discussion above). Now replace this copy of 1 by the entire sequence α and consolidate the resulting lists to form the list γ.

This somewhat complex discussion turns out not to be very significant in the definition of the $\{\circ_i\}$ above, since in that case all of the $\alpha_{t,i}$ 1's get replaced by lists (derived from the $s_{u,v}$'s and there is no consolidation of the result.

We are now in a position to prove the main result of this section:

THEOREM 2.30. *Let C be a weakly-coherent m-coalgebra, with $C_0 = \mathbb{Z}$ and $C_i = 0$ for $i < k$, where $k \geq 2$. Then $\mathcal{F}(C)$ has a natural structure as a strict weakly-coherent m-Hopf-algebra.*

---

[2] It is *unique* (so the word largest is unnecessary), if all of the β_j are positive.

Remarks.   2.30.1. See 3.13 on page 26 for a definition of m-Hopf algebra.

2.30.2. The construction given in the proof produces an m-structure that commutes with the product-operation of $\mathcal{F}(C)$. In the present context, this turns out to be equivalent to the requirement that the maps $h_n \colon \mathrm{R}(S_n) \to \mathcal{F}(\mathfrak{R}[C], n)$ be coalgebra homomorphisms. If we only want a weakly coherent m-structure on $\mathcal{F}(C)$, we can simply use the acyclicity of the $\{\mathcal{F}(\mathfrak{R}[C], n)\}$ to construct chain-maps $h_n \colon \mathrm{R}(S_n) \to \mathcal{F}(\mathfrak{R}[C], n)$.

PROOF. Most of the statement has already been proved — see 2.21 on page 50, 2.29 on page 56, and 2.28 on page 56. It only remains to verify the properties of a weakly coherent m-coalgebra — see 3.3 on page 19.

The composition operations defined in 2.29 on page 56 make the $\{\mathcal{F}(\mathfrak{R}[C], n)\}$ components of a formal coalgebra. The fact that formal composition in $\{\mathcal{F}(\mathfrak{R}[C], n)\}$ is compatible with composition of higher-coproducts of C (see 3.3 on page 19) follows from the fact that $(C, \{\mathfrak{f}[C]_n \colon C \to \mathrm{Hom}_{\mathbb{Z} S_n}(\mathfrak{R}[C]_n, C^n)\})$ was weakly coherent, definition 2.29 on page 56, and definition 2.24 on page 52. $\square$

Note that a number of arbitrary choices were made in each step of the derivation of the m-structure of $\mathcal{F}(C)$ in 2.30. We claim that these have no essential effect on the result. More precisely:

PROPOSITION 2.31. *Let* $\{f, \{h_n\}\} \colon (C, \{\mathfrak{f}[C]_n \colon C \to \mathrm{Hom}_{\mathbb{Z} S_n}(\mathfrak{R}[C]_n, C^n)\}) \to (D, \{\mathfrak{f}[D]_n \colon D \to \mathrm{Hom}_{\mathbb{Z} S_n}(\mathfrak{R}[D]_n, D^n)\})$ *be a strict morphism of weakly-coherent m-coalgebras. Then there is an induced map* $\mathcal{F}(\{f, \{g_n\}\}) \colon \mathcal{F}(C) \to \mathcal{F}(D)$ *that is also a strict morphism of weakly-coherent m-structures.*

PROOF. This follows immediately from the definitions in question. $\square$

The following result is well-known — we will need it to evaluate the effect of the cobar construction on a morphism of weakly-coherent m-coalgebras:

Recall 2.19 on page 49.

PROPOSITION 2.32. *Let* $(p, i, \varphi) \colon (C, \{\mathfrak{f}[C]_n \colon C \to \mathrm{Hom}_{\mathbb{Z} S_n}(\mathfrak{R}[C]_n, C^n)\}) \to (D, \{\mathfrak{f}[D]_n \colon D \to \mathrm{Hom}_{\mathbb{Z} S_n}(\mathfrak{R}[D]_n, D^n)\})\})$ *be an elementary morphism of weakly-coherent m-coalgebras, as defined in 3.9 on page 25. In addition suppose that* $C_j = 0$ *for* $j < 2$. *Then* $(\mathcal{F}(i), \mathcal{F}(p), \mathcal{F}(\varphi)) \colon \mathcal{F}(C) \to \mathcal{F}(D)$ *is an elementary morphism of the cobar constructions, where:*

1. $\mathcal{F}(i) = T(\downarrow \circ i \circ \uparrow) \colon \mathcal{F}(D) \to \mathcal{F}(C)$;
2. $\mathcal{F}(p) = T(\downarrow \circ p \circ \uparrow) \circ (1 + t \circ \mathcal{F}_\infty \circ T(\downarrow \circ \varphi \circ \uparrow)) \colon \mathcal{F}(C) \to \mathcal{F}(D)$;
3. $\mathcal{F}(\varphi) = \mathcal{F}_\infty \circ T(\downarrow \circ \varphi \circ \uparrow)$; *and* $t = \partial_{\mathcal{F}(C)} + T(\downarrow \circ \partial_C \circ \uparrow)$

$$\mathcal{F}_\infty = \sum_{i=1}^{\infty} (T(\downarrow \circ \varphi \circ \uparrow) \circ t)^i \colon \mathcal{F}(C) \to \mathcal{F}(C)$$

In addition, the map $\mathcal{F}(p)$ *is a homomorphism of DGA-algebras.*

Remarks. 2.32.1. Here $T(f)$ represents the induced map on the tensor-algebra. We define

$$T(\downarrow \circ \partial_C \circ \uparrow) = - \downarrow \circ \partial_C \circ \uparrow \oplus - \downarrow \circ \partial_C \circ \uparrow \otimes 1 \oplus -1 \otimes \downarrow \circ \partial_C \circ \uparrow \oplus$$
$$- \downarrow \circ \partial_C \circ \uparrow \otimes 1 \otimes 1 \oplus -1 \otimes \downarrow \circ \partial_C \circ \uparrow \otimes 1 \oplus -1 \otimes 1 \otimes \downarrow \circ \partial_C \circ \uparrow \ldots$$

and

$$T(\downarrow \circ \varphi \circ \uparrow) = \downarrow \circ \varphi \circ \uparrow \oplus 1 \otimes \downarrow \circ \varphi \circ \uparrow \oplus \downarrow \circ \varphi \circ \uparrow \otimes \downarrow \circ i \circ p \circ \uparrow$$
$$\oplus 1 \otimes 1 \otimes \downarrow \circ \varphi \circ \uparrow \oplus 1 \otimes \downarrow \circ \varphi \circ \uparrow \otimes \downarrow \circ i \circ p \circ \uparrow$$
$$\oplus \downarrow \circ \varphi \circ \uparrow \otimes \downarrow \circ i \circ p \circ \uparrow \otimes \downarrow \circ i \circ p \circ \uparrow \ldots$$

2.32.2. This result implies that morphisms, as defined in 3.9 and 3.10 on page 25, behave well with respect to the process of forming cobar constructions. Throughout the remainder of this paper we will define $\mathcal{F}(p, i, \varphi)$ to be $(\mathcal{F}(i), \mathcal{F}(p), \mathcal{F}(\varphi))$, as given above.

PROOF. The first statement is a straightforward application of the Perturbation Lemma, 2.19 (on page 49), where we regard the difference between the differential on the cobar construction and the tensor algebra of the de-suspension as the perturbation. We need only:

1. verify the hypotheses of 2.19 on page 49;
2. show that the map $\mathcal{F}(i)$ remains unperturbed, and the perturbed differential on $\mathcal{F}(D)$ is nothing but $\partial \mathcal{F}(D)$.

We define the filtration degree of $\Sigma^{-1} c_1 \otimes \cdots \otimes \Sigma^{-1} c_k$ to be $\sum_{i=1}^{k} \dim(\Sigma^{-1} c_i) - k$ or $\sum_{i=1}^{k} \dim(c_i) - 2k$. The requirement that $C_j = 0$ for $j < 2$ implies that $t(F_0 C) = 0$.

The fact that $\mathcal{F}(i)$ remains unperturbed follows immediately from the fact that the map i is a strict morphism of m-structures — it factors through the perturbation, t, giving the corresponding "perturbations" for the differential of the cobar construction of D. The additional perturbations to this differential (from 2.19 on page 49) vanish due to the annihilation properties of the maps in a contraction.

The last statement, regarding the fact that $\mathcal{F}(p)$ is a homomorphism of DGA-algebras follows by direct computation and the description of $T(\downarrow \circ \partial C \circ \uparrow)$ and $T(\downarrow \circ \varphi \circ \uparrow)$ in remark 2.19.1. It suffices to prove that $\mathcal{F}(i) \circ \mathcal{F}(p)$ is a homomorphism of DGA-algebras, since $\mathcal{F}(i) = T(\downarrow \circ i \circ \uparrow)$ is.

The fact that the induced differential on $\mathcal{F}(D)$ is that of the cobar construction follows from the fact that $\mathcal{F}(i)$ is injective, and it is unperturbed. The fact that it is injective implies that the induced differential on $\mathcal{F}(D)$ is the pullback of the differential on its image in $\mathcal{F}(C)$. The fact that $\mathcal{F}(i)$ is unperturbed implies that this coincides with the ordinary differential of the cobar construction. $\square$

COROLLARY 2.33. *Let* $g : (C, \{f[C]_n : C \rightarrow \mathrm{Hom}_{\mathbb{Z}S_n}(\mathfrak{R}[C]_n, C^n)\}) \rightarrow$ $(D, \{f[D]_n : D \rightarrow \mathrm{Hom}_{\mathbb{Z}S_n}(\mathfrak{R}[D]_n, D^n)\})$ *be a morphism of weakly-coherent m-coalgebras. Then there is an induced map* $\mathcal{F}(g) : \mathcal{F}(C) \rightarrow \mathcal{F}(D)$ *that is also a morphism of m-Hopf algebras.*

Remarks.   2.33.1. See 3.13 on page 26 for a definition of m-Hopf algebras.

2.33.2. In fact, the cobar-construction defines a functor from the category of 2-connected weakly-coherent m-coalgebras to the category of free m-Hopf algebras — see [17] and [14].

3. The Bar Construction

In this section we will compute weakly-coherent m-structures on the bar-construction. The bar construction was originally defined by Eilenberg and MacLane in [7] in 1954 — it predates the cobar construction. Like the cobar construction, it has a geometric significance:

> Suppose X is a simplicial group, and let $C(X)$ is its simplicial chain-complex. It is equipped with the structure of a DGA-algebra. The bar construction $\bar{\mathcal{B}}(C(X))$ is chain-homotopy equivalent to the chain-complex of the classifying space of X.

In [7] Eilenberg and MacLane proved that if π is an abelian group, then $\bar{B}^n(\mathbb{Z}\pi)$ is chain-homotopy equivalent to the chain-complex of the Eilenberg-MacLane space $K(\pi, n)$.

We begin with the reduced and unreduced bar construction:

PROPOSITION 3.1. *Suppose* $(C, \{\mathfrak{f}[C]_n \colon C \to \operatorname{Hom}_{\mathbb{Z}S_n}(\mathfrak{R}[C]_n, C^n)\})$ *is a weakly coherent m-Hopf algebra and we equip each of the chain-modules of* $\{\mathfrak{R}[C]_n\}$ *with a basis. Then the reduced and unreduced bar constructions* $\bar{\mathcal{B}}(C)$ *and* $\mathcal{B}(C)$ *are weakly coherent m-Hopf algebras of the form* $(\bar{\mathcal{B}}(C), \{\mathfrak{f}[\bar{\mathcal{B}}(C)]_n \colon \bar{\mathcal{B}}(C) \to \operatorname{Hom}_{\mathbb{Z}S_n}(\mathfrak{R}[C]_n, \bar{\mathcal{B}}(C)^n)\})$ *and* $(\mathcal{B}(C), \{\mathfrak{f}[\mathcal{B}(C)]_n \colon \mathcal{B}(C) \to \operatorname{Hom}_{\mathbb{Z}S_n}(\mathfrak{R}[C]_n, \mathcal{B}(C)^n)\})$. *If* $(C, \{\mathfrak{f}_n \colon C \to \operatorname{Hom}_{\mathbb{Z}S_n}(\mathfrak{R}[C]_n, C^n)\})$ *is strongly coherent, then so are* $(\bar{\mathcal{B}}(C), \{\mathfrak{f}[\bar{\mathcal{B}}]_n \colon \bar{\mathcal{B}}(C) \to \operatorname{Hom}_{\mathbb{Z}S_n}(\mathfrak{R}[C]_n, \bar{\mathcal{B}}(C)^n)\})$ *and* $(\mathcal{B}(C), \{\mathfrak{f}[\mathcal{B}(C)] \colon \mathcal{B}(C) \to \operatorname{Hom}_{\mathbb{Z}S_n}(\mathfrak{R}[C]_n, \mathcal{B}(C)^n)\})$.

Remarks.   3.1.1. See 3.13 on page 26 for a definition on an m-Hopf algebra.

3.1.2. Recall the definition of an f-resolution in 3.1 on page 19. We construct the m-structure on $\mathcal{B}(C)$ as follows:

Suppose $b = [c_1| \ldots |c_k] \otimes c_{k+1} \in \bar{\mathcal{B}}(C) \otimes_x C$.   The images of $\mathfrak{f}[C]_n(c_1), \ldots, \mathfrak{f}[C]_n(c_{k+1})$ lie in maps from $\mathfrak{R}[C]_n \to C^n$ that factor through finitely generated $\mathbb{Z}S_n$-chain complexes $M_1, \ldots, M_{k+1}$, respectively. Let M be the push-out of the surjections $\mathfrak{R}[C]_n \to M_i$, where $i = 1, \ldots, k + 1$ — this is finitely generated because the group-ring $\mathbb{Z}S_n$ is *noetherian*.

The proof of this result is exactly like that of the proof of coherency of $\mathcal{C}(*)$ on page 29: We will define $\mathfrak{f}[\mathcal{B}(C)]([c_1| \ldots |c_k] \otimes c_{k+1})$ to be a map $M \to \mathcal{B}(C)^n$. This map is constructed inductively via

$$\mathfrak{f}[\mathcal{B}(C)]_n(b \otimes 1)(\sigma)$$
$$= H_n \circ \left\{ \mathfrak{f}[\mathcal{B}(C)](\partial_{\bar{\mathcal{B}}(C) \otimes_x C}(b \otimes 1))(\sigma) + (-1)^{\dim(b)} \mathfrak{f}[\mathcal{B}(C)](b \otimes 1)(\partial \sigma) \right\}$$

— this formula requires σ to be a canonical basis element of M — i.e., a basis element that is the image of a preferred generating element from $\mathfrak{R}[C]_n$. Here $H_n = 1 \otimes \cdots \otimes$

$1 \otimes S + 1 \otimes \cdots \otimes 1 \otimes S \otimes \epsilon + \cdots$ — where $S([c_1|\ldots|c_k] \otimes c_{k+1}) = [c_1|\ldots|c_k|c_{k+1}]$, and, as usual, ϵ is the augmentation map. We extend the definition of $\mathcal{B}(f_n)(c)$ to all of $\mathfrak{R}[C]_n$ by defining it to be $\mathbb{Z}S_n$-linear. Since $\bar{\mathcal{B}}(C) \otimes_x C$ is a twisted tensor product the boundary $\partial_{\bar{\mathcal{B}}(C) \otimes_x C}(b \otimes 1)$ will generally contain a nontrivial factor of C — we consequently use the lower dimensional formula for $\mathfrak{f}[\mathcal{B}(C)]_n(b' \otimes 1)(\sigma)$, coupled with the formula $\mathfrak{f}\widetilde{[\mathcal{B}(C)]} = \mu \circ (\mathfrak{f}\widetilde{[\mathcal{B}(C)]}(* \otimes 1) \otimes \mathfrak{f}\widetilde{[C]}(*)) \circ \Delta_{\mathfrak{R}[C]} \otimes 1$ to compute $\mathfrak{f}[\mathcal{B}(C)](\partial_\otimes(b \otimes 1))(\sigma)$ and then we apply the chain-homotopy H_n. Note that the map H_n is not $\mathbb{Z}S_n$-linear.

It is not hard to see that this defines an element of $\mathrm{Hom}_{\mathbb{Z}S_n}(\mathfrak{R}[C]_n, \mathcal{B}(C)^n)$. It is only necessary to verify that, if we had another finitely-generated chain complex, M'', such that the surjection $\mathfrak{R}[C]_n \to M$ factored through M'', then the maps constructed by the algorithm given above would fit in a commutative diagram

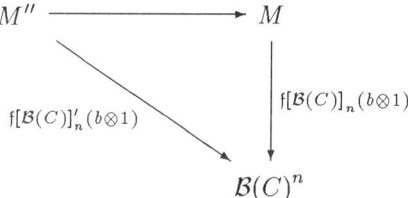

This follows from the formula used above and the fact that the basis elements of M'' and M used to perform the computations must be images of preferred generating elements of $\mathfrak{R}[C]_n$, hence must be compatible.

This completes the construction of the m-structure on $\mathcal{B}(C)$. This induces a well-defined weakly coherent m-structure on $\bar{\mathcal{B}}(C)$ as well.

It remains to show that the bar-construction is an m-Hopf algebra, in the sense of 3.13 on page 26. We construct a weakly-coherent m-coalgebra structure on the algebraic mapping cylinder of the (shuffle) product-mapping $\mu: \mathcal{B}(C) \otimes \mathcal{B}(C) \to \mathcal{B}(C)$, using an argument like that used in the proof of 4.1 on page 28 and 4.5 on page 31.

3.1.3. The m-structures of $\bar{\mathcal{B}}(C)$ and $\mathcal{B}(C)$ depend upon the bases chosen. It is clear that the bar constructions (as defined above) carries strict morphisms that preserve canonical bases into strict morphisms.

3.1.4. This implies that the DGA-algebras $A(M, n)$, defined in [7] are coherent m-Hopf algebras.

PROPOSITION 3.2. *Let $\{b_1\}$ and $\{b_2\}$ be two distinct equivalent sets of canonical bases on the coordinate coalgebras $\{\mathfrak{R}[C]_n\}$ of a weakly-coherent m-coalgebra $(C, \{\mathfrak{f}[C]_n \colon C \to \mathrm{Hom}_{\mathbb{Z}S_n}(\mathfrak{R}[C]_n, C^n)\})$. Then there exists an equivalence between $(\bar{\mathcal{B}}(C), \{\mathfrak{f}[\bar{\mathcal{B}}(C)]_n, \{b_1\}) \colon \bar{\mathcal{B}}(C) \to \mathrm{Hom}_{\mathbb{Z}S_n}(\mathfrak{R}[C]_n, \bar{\mathcal{B}}(C)^n)\})$ and $(\bar{\mathcal{B}}(C), \mathfrak{f}[\bar{\mathcal{B}}(C)]_n, \{b_2\}) \colon \bar{\mathcal{B}}(C) \to \mathrm{Hom}_{\mathbb{Z}S_n}(\mathfrak{R}[C]_n, \bar{\mathcal{B}}(C)^n)\})$.*

REMARK. 3.2.1. This result is necessary because the construction of the derived co-ordinates maps, the $\{\mathfrak{f}[\bar{\mathcal{B}}(C)]_n\}$, depends explicitly upon the choice of canonical basis of the coordinate rings.

PROOF. Form the chain-complex and a multiplicative construction (in the sense of [5]) $\mathfrak{R}[C]_n \otimes \mathrm{R}(S_n) \otimes \bar{\mathcal{B}}(C) \otimes_x C \otimes I$. This is nothing but the $\mathbb{Z}$-tensor product of the m-coalgebras $\bar{\mathcal{B}}(C) \otimes_x C$ and I (regarding this as a 1-simplex and using $\mathcal{C}(\sigma^1)$). We define the adjoint coordinate maps for $\bar{\mathcal{B}}(C) \otimes_x C \otimes I$ and we compute an m-structure on it as follows: on

$$\mathfrak{R}[C]_n \otimes \mathrm{R}(S_n) \otimes \bar{\mathcal{B}}(C) \otimes C \otimes \{0\}$$

we use the $\{b_1\}$ as canonical basis elements for the $\{\mathfrak{R}[C]_n\}$; on $\mathfrak{R}[C]_n \otimes \mathrm{R}(S_n) \otimes \bar{\mathcal{B}}(C) \otimes C \otimes \{1\}$ and $\mathfrak{R}[C]_n \otimes \mathrm{R}(S_n) \otimes \bar{\mathcal{B}}(C) \otimes C \otimes q$, where q is the 1-dimensional simplex of the unit interval, we use $\{b_2\}$. In all cases, we use tensor products of basis elements as canonical basis elements of $\mathfrak{R}[C]_n \otimes \mathrm{R}(S_n)$. The result is an m-structure on $\bar{\mathcal{B}}(C) \otimes_x C \otimes I$ with the property that $(\bar{\mathcal{B}}(C), \{\mathfrak{f}[\bar{\mathcal{B}}(C)]_n, \{b_1\}) : \bar{\mathcal{B}}(C) \to \mathrm{Hom}_{\mathbb{Z}S_n}(\mathfrak{R}[C]_n, \bar{\mathcal{B}}(C)^n)\})$ and $(\bar{\mathcal{B}}(C), \mathfrak{f}[\bar{\mathcal{B}}(C)]_n, \{b_2\}) : \bar{\mathcal{B}}(C) \to \mathrm{Hom}_{\mathbb{Z}S_n}(\mathfrak{R}[C]_n, \bar{\mathcal{B}}(C)^n)\})$. map to the ends via strict morphisms. This proves the result. $\square$

COROLLARY 3.3. *Let*

$$f : (C, \{\mathfrak{f}[C]_n : C \to \mathrm{Hom}_{\mathbb{Z}S_n}(\mathfrak{R}[C]_n, C^n)\})$$
$$\to (D, \{\mathfrak{f}[D]_n : D \to \mathrm{Hom}_{\mathbb{Z}S_n}(\mathfrak{R}[D]_n, D^n)\})$$

be a strict morphism of weakly-coherent m-coalgebras. Then f induces a morphism of bar-constructions

$$\bar{\mathcal{B}}(f) : (\bar{\mathcal{B}}(C), \{\mathfrak{f}[C]_n(\bar{\mathcal{B}}) : \bar{\mathcal{B}}(C) \to \mathrm{Hom}_{\mathbb{Z}S_n}(\mathfrak{R}[C]_n, \bar{\mathcal{B}}(C)^n)\})$$
$$\to (\bar{\mathcal{B}}(D), \{\mathfrak{f}[D]_n(\bar{\mathcal{B}}) : \bar{\mathcal{B}}(D) \to \mathrm{Hom}_{\mathbb{Z}S_n}(\mathfrak{R}[D]_n, \bar{\mathcal{B}}(D)^n)\})$$

PROOF. This is very similar to the proof of 3.2 — when forming $\mathfrak{R}[C]_n \otimes \mathfrak{R}[C]_n \otimes (\bar{\mathcal{B}}(C), \{\mathfrak{f}[\bar{\mathcal{B}}(C)]_n : \bar{\mathcal{B}}(C) \to \mathrm{Hom}_{\mathbb{Z}S_n}(\mathfrak{R}[C]_n, \bar{\mathcal{B}}(C)^n)\}) \otimes I$ we form the algebraic mapping cylinder of $\bar{\mathcal{B}}(f) \otimes f$ (which is perfectly well-defined independently of the question of the existence of an m-structure). This amounts to redefining the boundary of $\bar{\mathcal{B}}(C) \otimes C \otimes I$ so that $\bar{\mathcal{B}}(C) \otimes C \otimes \{1\}$ becomes replaced by $\bar{\mathcal{B}}(D) \otimes D \otimes \{1\}$ and the boundary acts like $\bar{\mathcal{B}}(f) \otimes f$. Note that the fact that the coordinate map $g : \mathfrak{R}[D]_n \to \mathfrak{R}[C]_n$ might not be a chain-map presents no problems:

1. It commutes with the structure maps f_n, and $\mathfrak{f}[D]_n$;
2. It must be a DGA-coalgebra homomorphism (see remark 3.13.2 on page 27).

$\square$

COROLLARY 3.4. *Let*

$$f : (C, \{\mathfrak{f}[C]_n : C \to \mathrm{Hom}_{\mathbb{Z}S_n}(\mathfrak{R}[C]_n, C^n)\})$$
$$\to (D, \{\mathfrak{f}[D]_n : D \to \mathrm{Hom}_{\mathbb{Z}S_n}(\mathfrak{R}[D]_n, D^n)\})$$

be a morphism of weakly-coherent m-Hopf-algebras. Then f induces a morphism of bar-constructions

$$\bar{\mathcal{B}}(f) : (\bar{\mathcal{B}}(C), \{\mathfrak{f}[\bar{\mathcal{B}}(C)]_n : \bar{\mathcal{B}}(C) \to \mathrm{Hom}_{\mathbb{Z}S_n}(\mathfrak{R}_n, \bar{\mathcal{B}}(C)^n)\})$$
$$\to (\bar{\mathcal{B}}(D), \{\mathfrak{f}[D]_n(\bar{\mathcal{B}}) : \bar{\mathcal{B}}(D) \to \mathrm{Hom}_{\mathbb{Z}S_n}(\mathfrak{R}[D]_n, \bar{\mathcal{B}}(D)^n)\})$$

REMARK. 3.4.1. We simply form bar-constructions of the algebraic mapping cylinders that occur in the structure of f. We need not be concerned with questions of canonical bases for the coordinate coalgebras because of 3.2 on page 61.

We can now use the Cartan Theory of Constructions (in [5]) to compute a product-operation on the bar construction. With this product the bar construction becomes a weakly-coherent m-Hopf algebra.

PROPOSITION 3.5. *The derived m-structure on the bar construction is geometrically valid. In other words, if X is a pointed simply-connected CW complex and $g: \bar{\mathcal{B}}(X) \to \bar{W}X$ is the natural map defined by Eilenberg and MacLane in [7], then g is a morphism of m-coalgebras $g: \bar{\mathcal{B}}(\mathcal{C}(X)) \to \mathcal{C}(\bar{W}X)$.*

REMARK. 3.5.1. This is proved in appendix B on page 103.

We will conclude this section by considering an important special case.

PROPOSITION 3.6. *Let $[g_1| \ldots |g_k] \in \mathrm{R}(S_n)$ and let $c \in C$. Then*

$$f[\bar{\mathcal{B}}(C)]_n([c])([g_1| \ldots |g_k])$$

$$= \sum_{i=1}^{\min(k,n-1)} (-1)^{i \cdot (\dim(c)+1)} p^n \circ H_n \circ g_1 \circ \cdots \circ g_i \circ H_n$$

$$\circ f[C]_n(c)([g_{i+1}| \ldots |g_k])$$

Remarks. 3.6.1. We will need to know the m-structure on these elements of $\bar{\mathcal{B}}(C)$ in the sequel. The map p is the projection $\mathcal{B}(C) \to \bar{\mathcal{B}}(C)$. It is essentially characterized by the properties: $p \circ S = S$, $p \circ \epsilon = \epsilon$, $p(u \otimes c) = 0$, where u is an arbitrary element of $\bar{\mathcal{B}}(C)$, and $c \in C$ is of dimension > 0.

3.6.2. We can re-write the term $H_n \circ g_1 \circ \cdots \circ g_i \circ H_n$ as: $g_1 \ldots \cdot g_i \circ H_n^{g_i^{-1} \ldots g_1^{-1}} \circ H_n^{g_i^{-1} \ldots g_2^{-1}} \circ \cdots \circ H_n^{g_i^{-1}} \circ H_n$, where $H_n^\sigma = \sigma \circ H_n \circ \sigma^{-1}$ is the result of permuting the factors of the homotopy, H_n, via σ. We assume the Koszul conventions apply when we compose permuted H_n-functions. This re-writing process is significant because it implies that a given term of the sum above will vanish if any two of the products $\{g_1 \ldots g_i, g_2 \ldots g_i, \ldots, g_i\}$ are equal. This is due to the fact that the homotopy, S, is self-annihilating. The equality of any two of the group elements above implies that for some $1 \leq i' \leq i'' < i$, $g_{i'} \ldots g_{i''} = 1$.

PROOF. Recall that the map H_n sends one factor of its argument to $\bar{\mathcal{B}}(C) \otimes 1 \subset \bar{\mathcal{B}}(C) \otimes_x C$. We use the formula $f[\bar{\mathcal{B}}(C)]_n([b] \otimes 1)(\sigma) = H_n \circ f[C]_n(b)(\sigma) + (-1)^{\dim([b])} H_n \circ f[\bar{\mathcal{B}}(C)]_n([b] \otimes 1)(\partial\sigma)$ from 3.1, recursively. Each term of this result consists in a string of factors that represent the effects of composing copies of the first and second terms of the formula.

Claim 1: An application of the first term is never substituted into another application of the first term. This is due to the fact that H_n is self-annihilating.

Claim 2: An application of the second term is never substituted into an application of the first term. This is due to the fact that H_n is self-annihilating.

Two successive copies of the second term can occur or applications of the first term can be substituted into the second term, however, since H_n is not $\mathbb{Z}S_n$-linear. This is what the formula in the statement of the results contains. □

Now we consider the trivial m-coalgebra:
Let $C_0 = \mathbb{Z}[M]$ where M is a $\mathbb{Z}$-module and $C_i = 0$ for $i > 0$

(3.1)
$$\mathfrak{f}[C]_n(g)(a) = \begin{cases} g \otimes \cdots \otimes g & (n \text{ factors}) \text{ if } a = [\,] \\ 0 & \text{if } \dim(a) > 0 \end{cases}$$

This is clearly a coherent m-coalgebra. Its existence and 3.5 on page 63 implies:

PROPOSITION 3.7. *If M is a $\mathbb{Z}$-module and n is an integer ≥ 1 then the iterated bar constructions $A(M, n)$ have the natural coherent m-coalgebra structure, that is equivalent (when $n > 1$) to the geometric m-coalgebra structures on $K(M, n)$.*

Let C be an arbitrary coherent m-Hopf algebra and consider the canonical acyclic twisted tensor product $\bar{\mathcal{B}}(C) \otimes_x C$ and the classifying map of its twisting cochain: $f(x) \colon \mathcal{F}(\bar{\mathcal{B}}(C)) \to C$. Recall the definition of simplicial dimension $\dim_S(b)$ of a canonical basis element $b = [c_1| \ldots |c_k] \in \bar{\mathcal{B}}(C)$ from [5]: it is equal to the number n. We now consider what the map $f(x)$ looks like: The twisting cochain $x \colon \bar{\mathcal{B}}(C) \to C$, vanishes on basis elements b with $\dim_S(b) > 1$, and on elements of the form $[c]$ it is essentially $\downarrow$ sending $[c]$ to $c \in C$. It follows that:

1. $f(x)$ vanishes on any tensor product of basis elements, $b_1 \otimes \otimes b_k$, with $\dim_S(b_i) > 1$ for any i or with any factor equal to $[\,]$ (the identity element of $\bar{\mathcal{B}}(C)$);
2. $f(x)([c_1] \otimes \cdots \otimes [c_k]) = c_1 \cdots c_k \in C$.

In the formula for H_n (in 3.6 on page 63), we can ignore all terms but the first — i.e. H_n is effectively equal to $1 \otimes \cdots \otimes 1 \otimes S$ — see 3.6 and the discussion preceding it for the notation. This is due to the fact that all factors of the f_n-functions of $\bar{\mathcal{B}}(C)$ will be plugged into $\downarrow^{|\alpha|}$ for some value of $\alpha > 0$, and $\downarrow$ annihilates any element of dimension 0 — see 2.8 on page 42 and the remarks following it.

There exists an inclusion of graded modules $\Sigma C \to\!\!\!\to \bar{\mathcal{B}}(C)$ and this induces an inclusion $\mathfrak{J} \colon C \to \mathcal{F}(\bar{\mathcal{B}}(C))$ that maps $c \in C$ to $\Sigma^{-1}[c] \in \mathcal{F}(\bar{\mathcal{B}}(C))$. The definition of the coproduct of $\bar{\mathcal{B}}(C)$ in [8] implies that the map $\mathfrak{J}$ is a chain-map (and even a chain-homotopy equivalence). This is due to the fact that the coproduct of $[c]$ is $1 \otimes [c] + [c] \otimes 1$, which vanishes when projected to $\Sigma^{-1}\bar{\mathcal{B}}(C)^+ \otimes \Sigma^{-1}\bar{\mathcal{B}}(C)^+ \subset \mathcal{F}(\bar{\mathcal{B}}(C)) \otimes \mathcal{F}(\bar{\mathcal{B}}(C))$.

LEMMA 3.8.   1. *Let $\phi_{\mathfrak{J}} \colon \mathcal{F}(\bar{\mathcal{B}}(C)) \to \mathcal{F}(\bar{\mathcal{B}}(C))$ be defined inductively by:*
   a. $\phi_{\mathfrak{J}}|\Sigma^{-1}\bar{\mathcal{B}}(C) \subset \mathcal{F}(\bar{\mathcal{B}}(C)) = 0$;
   b. $\phi_{\mathfrak{J}}(\Sigma^{-1}[c] \otimes \Sigma^{-1}[c_1| \ldots |c_k]) = (-1)^{\dim(c)}\Sigma^{-1}[c|c_1| \ldots |c_k] \in \Sigma^{-1}\bar{\mathcal{B}}(C) \subset \mathcal{F}(\bar{\mathcal{B}}(C))$, *where $c, c_1, \ldots, c_k \in C$;*
   c. $\phi_{\mathfrak{J}}(b_1 \otimes b_2 \otimes g) = \phi_{\mathfrak{J}}(b_1 \otimes b_2) \otimes g + \phi_{\mathfrak{J}}((\mathfrak{J} \circ f(x))(b_1 \otimes b_2) \otimes g)$, *where $g \in \mathcal{F}(\bar{\mathcal{B}}(C)), b_1, b_2 \in \Sigma^{-1}\bar{\mathcal{B}}(C)$. Then the triple $(f(x), \mathfrak{J}, \phi_{\mathfrak{J}}) \colon \mathcal{F}(\bar{\mathcal{B}}(C)) \to C$, constitutes a contraction of DGA-algebras.*

DEFINITION 3.9. Let $K_n = \Sigma^{-(n-1)} R(S_n)^+$ as a chain complex to but with the action of $\mathbb{Z}S_n$ induced by regarding it as a subcomplex of $\Sigma Z_{1,\ldots,1} \subset \bar{\mathcal{B}}(Z_{n,m}(R(S_n)))$ — see remark 2.7.2 on page 42. Define the mapping $\mathfrak{M}'_n \colon K_n \to R(S_n)$ as follows:

1. It is defined on a canonical basis element $[g_1|\ldots|g_k]$, with $k \geq n-1$ and $g_1,\ldots,g_k \in S_n$, via $\mathfrak{M}'_n([g_1|\ldots|g_k]) = (-1)^{n-1} g_1 \cdots g_{n-1} \mathfrak{P}(g_1,\ldots,g_{n-1})[g_n|\ldots|g_k]$, where $[g_n|g_{n-1}]$ is regarded as $[\,]$, and $\mathfrak{P}(g_1,\ldots,g_{n-1})$ is defined to be:
   a. 0, if the numbers $\{g_{n-1}^{-1} \circ \cdots \circ g_1^{-1}(n),\ldots,g_{n-1}^{-1} \circ g_{n-2}^{-1}(n), g_{n-1}^{-1}(n), n\}$ are not all distinct;
   b. $\mathrm{parity}(\wp(g_1,\ldots,g_{n-1}))$ times the parity of the product $(g_1 \cdots g_{n-1})$, where $\wp(g_1,\ldots,g_{n-1})$ denotes the permutation that maps the sequence $\{1,\ldots,n\}$ into the sequence $\{g_{n-1}^{-1} \circ \cdots \circ g_1^{-1}(n),\ldots,g_{n-1}^{-1} \circ g_{n-2}^{-1}(n), g_{n-1}^{-1}(n), n\}$ (if they are all distinct);
2. $\mathfrak{M}'_n$ is a $\mathbb{Z}S_n$-linear homomorphism of graded modules.

REMARK. 3.9.1. $\Sigma^{-(n-1)} R(S_n)^+$ represents the result of truncating $\Sigma^{-(n-1)} R(S_n)$ in dimension 0. If we regard $\mathfrak{M}'_n$ as a map $\Sigma^{-(n-1)}R(S_n)^+ \to R(S_n)$, then it is a twisted $\mathbb{Z}S_n$-homomorphism in the sense that: $\mathfrak{M}'_n(g \cdot a) = (-1)^{\mathrm{parity}(g)} g \cdot \mathfrak{M}'_n(a)$, where $g \in S_n, a \in R(S_n)$.

Statement 2 of 2.29 on page 56 (with $k = 1$ and $a_1 = \{1,\ldots,1\}$) and the results of appendix A (particularly 2.17 on page 17) imply that:

COROLLARY 3.10. *We can define a morphism of formal coalgebras* $\mathfrak{M}_n \colon \mathcal{F}(R(S_n), n) \to R(S_n)$ *by:*

1. *mapping* $\mathcal{F}(R(S_n), n)$ *to* $\mathcal{F}(R(S_n), n)/Z_{n,m}$ *(see 2.26 on page 55);*
2. *defining* $\mathfrak{M}_n = \mathfrak{M}'_n|S$ *(see 2.13)* $= \Sigma^{-(n-1)} R(S_n)^+ \subset \bar{\mathcal{B}}(()Z_{n,m}(R(S_n)))$ — *this is a direct summand of* $\bar{\mathcal{B}}(Z_{n,m}(R(S_n)))$, *as a graded module;*
3. *defining it to be zero on all other summands* $Z_\alpha(R(S_n))$ *and all elements of* $\bar{\mathcal{B}}(Z_{n,m}(R(S_n)))$ *of the form* $[g_1|\ldots|g_k]$, *with* $k > 1$.

REMARK. 3.10.1. This is proved in appendix A, where A.3 on page 100 proves that it is a chain-map and 2.17 on page 17 proves that it induces a morphism of formal coalgebras.

The results of 3.9 on page 65 imply that:

PROPOSITION 3.11. *The following diagram commutes for all values of* n:

$$\mathrm{Hom}_{\mathbb{Z}S_n}(R(S_n), C^n) \xrightarrow{\mathrm{Hom}_{\mathbb{Z}S_n}(\mathfrak{M}_n, \mathfrak{J}^n)} \mathrm{Hom}_{\mathbb{Z}S_n}(\mathcal{F}(R(S_n), n), \mathcal{F}(\bar{\mathcal{B}}(C))^n)$$

$$\Big\uparrow {\scriptstyle \mathfrak{f}_n} \qquad\qquad\qquad\qquad\qquad\qquad\qquad \Big\uparrow {\scriptstyle \mathcal{F}(\mathfrak{f}[\bar{\mathcal{B}}(C)]_n)}$$

$$C \xrightarrow{\quad \mathfrak{J} \quad} \mathcal{F}(\bar{\mathcal{B}}(C))$$

Consequently, $(\mathfrak{I}, \mathfrak{M}_n): C \rightarrow \mathcal{F}(\bar{\mathcal{B}}(C))$, *is a strict morphism of m-coalgebras, and* $(f(x), \mathfrak{I}, \phi_{\mathfrak{I}}): \mathcal{F}(\bar{\mathcal{B}}(C)) \rightarrow C$, *is a morphism of m-coalgebras that is also a homomorphism of DGA-algebras.*

4. Geometricity of the m-structure on the cobar construction

We conclude this paper with a proof of the geometricity of the m-coalgebra structure given in the present paper for the cobar construction.

LEMMA 4.1. *Let* X *be a pointed simplicial group, equipped with the canonical m-structure defined in 4.2 on page 30. Consider the contraction* $(f, g, \varphi): \bar{W}(X) \rightarrow \bar{\mathcal{B}}(X)$, *defined in* [26]. *Then the map* $g: \bar{\mathcal{B}}(X) \rightarrow \bar{W}(X)$ *is a strict morphism of m-structures. Consequently, the map* $f: \bar{W}(X) \rightarrow \bar{\mathcal{B}}(X)$, *is a morphism of coherent m-coalgebras.*

REMARK. 4.1.1. This m-structure is derived from the canonical simplicial m-structure on X and is coherent. This lemma is proved in appendix B on page 103. The map of coordinate coalgebras in this case is the identity map.

THEOREM 4.2. *Let* X *be a pointed simplicial complex with loop space* ΩX *and suppose both chain-complexes are equipped with the coherent m-structures defined in § 4 on page 28. Then there exists a morphism of m-coalgebras* $u: \mathcal{F}(X) \rightarrow \Omega X$ *that is also a morphism of DGA-algebras and whose underlying map is a chain-homotopy equivalence.*

REMARK. 4.2.1. The corresponding coordinate map is $\mathfrak{M}_n$, defined in 3.9 on page 65.

PROOF. We make extensive use of the results of the end of § 3 on page 60 and appendix B on page 103. This map is the composite $\mathcal{F}(X) \rightarrow \mathcal{F}(\bar{\mathcal{B}}(\Omega X)) \rightarrow \Omega X$, where the map on the right is a morphism of m-coalgebras (see 3.8 on page 24 and 3.10 in that section) algebras and a chain-homotopy equivalence, since it is the type of map defined in 3.11 on page 65. The map on the left is $\mathcal{F}(r)$, where $r = b(c): X \rightarrow \bar{\mathcal{B}}(\Omega X)$, is the classifying map of the twisted tensor product $X \otimes_c \Omega X$ defined via the main results of [10] — this twisted tensor product is acyclic since it is chain-homotopy equivalent to the total space of the canonical fibration over X with fiber ΩX. The map, r, is consequently a DGA-coalgebra homomorphism. It turns out to also be a morphism of m-Hopf algebras because it is the composite of a strict morphism (see 3.8 on page 24) of m-Hopf algebras with the inverse of another such strict morphism (see 3.10 on page 25). Consider the diagram:

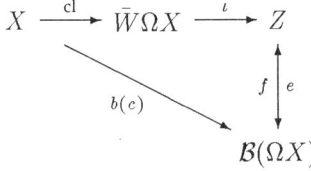

Here, cl is induced by the (geometric) classifying map of the path-space fibration $X \times_z \Omega X$, hence a strict morphism of m-structures and $Z = P(X)$ is as defined in B.6 on page 106. The diagram commutes in the clockwise direction, but not in the counter-clockwise direction. Proposition 3.11 on page 65 implies that this composite is a morphism. Consequently, the whole composite, u, is a morphism. Since it is the induced map of cobar constructions, it is a morphism of m-Hopf algebras, by 2.33 on page 59. □

COROLLARY 4.3. *Let X be an n-connected pointed simplicial complex equipped with the canonical (coherent) m-structure, $\mathcal{C}(X)$. If $i < n$ is an integer then there exists a morphism of m-coalgebras $\mathcal{F}^{i-1}(u)\colon \mathcal{F}^i\mathcal{C}(X) \to \mathcal{C}(\Omega^i X)$, that induces a homotopy equivalence of underlying chain-complexes and is a morphism of DGA-algebras.*

PROOF. This follows by an easy induction on i. The case where $i = 1$ is already handled by 4.2 on page 66. If the result has been proved for all $i < k$, we take the cobar construction of $\mathcal{F}^{k-2}(u)\colon \mathcal{F}^{k-1}\mathcal{C}(X) \to \mathcal{C}(\Omega^{k-1}X)$ to get $\mathcal{F}^{k-1}(u)\colon \mathcal{F}^k\mathcal{C}(X) \to \mathcal{F}\mathcal{C}(\Omega^{k-1}X)$ and compose this with $u(\Omega^{k-1}X)\colon \mathcal{F}\mathcal{C}(\Omega^{k-1}X) \to \mathcal{C}(\Omega^k X)$. This proves the result. □

COROLLARY 4.4. *Let X be a CW-complex and let $f_n\colon \mathcal{C}(X) \to \mathrm{Hom}_{\mathbb{Z} S_n}(\mathrm{R}(S_n), \mathcal{C}(X)^n)$ be the structure maps of its canonical m-structure. Then the morphism u defined in 4.2 on page 66 fits into the following homotopy-commutative diagram:*

$$\mathrm{Hom}_{\mathbb{Z} S_n}(\mathrm{R}(S_n), (\mathcal{F}\mathcal{C}(X))^n) \xrightarrow{\mathrm{Hom}_{\mathbb{Z} S_n}(1, u^n)} \mathrm{Hom}_{\mathbb{Z} S_n}(\mathcal{F}(\mathrm{R}(S_n), n), (\mathcal{C}(\Omega X))^n)$$

$$\mathcal{F}(f_n) \Big\uparrow \qquad\qquad\qquad\qquad\qquad \Big\uparrow \mathrm{Hom}_{\mathbb{Z} S_n}(\mathfrak{M}_n, 1)\circ f_n$$

$$\mathcal{F}(\mathcal{C}(X)) \xrightarrow{\quad\quad u \quad\quad} \mathcal{C}(\Omega X)$$

PROOF. This follows from 3.12 on page 25. □

Here, the target of the map is a non-coherent m-coalgebra. In order to determine the significance of this diagram, it is necessary to find a chain map $h_n\colon \mathrm{R}(S_n) \to \mathcal{F}(\mathrm{R}(S_n), n)$ that is right inverse to the map $\mathfrak{M}_n$, defined in 3.10 on page 65. This is done in appendix B on page 103 — the map in question is computed using the algorithm in 2.27 on page 55. We get:

COROLLARY 4.5. *Let X be a CW-complex and let $f_n\colon \mathcal{C}(X) \to \mathrm{Hom}_{\mathbb{Z} S_n}(\mathrm{R}(S_n), \mathcal{C}(X)^n)$ be the structure maps of its canonical m-structure. Then the following diagram is homotopy-commutative:*

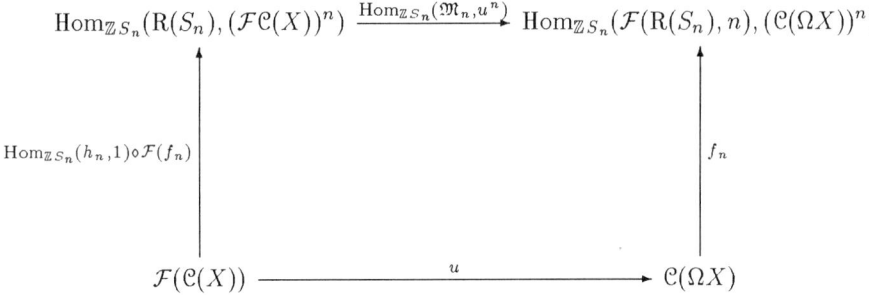

$$\mathrm{Hom}_{\mathbb{Z}S_n}(\mathrm{R}(S_n),(\mathcal{F}\mathcal{C}(X))^n) \xrightarrow{\mathrm{Hom}_{\mathbb{Z}S_n}(\mathfrak{M}_n,u^n)} \mathrm{Hom}_{\mathbb{Z}S_n}(\mathcal{F}(\mathrm{R}(S_n),n),(\mathcal{C}(\Omega X))^n)$$

$$\mathrm{Hom}_{\mathbb{Z}S_n}(h_n,1)\circ\mathcal{F}(f_n) \Big\uparrow \qquad\qquad\qquad\qquad \Big\uparrow f_n$$

$$\mathcal{F}(\mathcal{C}(X)) \xrightarrow{\quad u \quad} \mathcal{C}(\Omega X)$$

CHAPTER 4

Fibrations and twisted tensor products

1. The functors $Y_{n,m}(\mathfrak{R})$

DEFINITION 1.1. Let $(C,\ \{f[C]_n\colon C\ \to\ \mathrm{Hom}_{\mathbb{Z}S_n}(\mathfrak{R}[C]_n, C^n)\})$ be a weakly-coherent m-coalgebra. If $\alpha = \{\alpha_1, \dots, \alpha_n\}$ is a sequence of n integers, each ≥ 0 define $Y(\mathfrak{R}[C], \alpha) = \downarrow^{|\alpha|} \mathfrak{R}[C]_{n+|\alpha|}$, where $1 \leq i \leq n$, and

$$Y_{n,m}(\mathfrak{R}[C]) = \bigoplus_{\substack{\alpha_1,\dots,\alpha_n \\ |\alpha| \leq m}} Y(\mathfrak{R}[C], \alpha)$$

where the summation is taken over all i and all possible sequences α of length n (of nonnegative integers).

As with the $Z_{n,m}(\mathfrak{R}[C])$, the $Y_{n,m}(\mathfrak{R}[C])$ form an inverse system

$$q_{m',m}\colon Y_{n,m'}(\mathfrak{R}[C]) \to Y_{n,m}(\mathfrak{R}[C])$$

for all pairs m, m' such that $m' > m$.

REMARK. 1.1.1. The Y-complexes will turn out to be factors of the coordinate coalgebras that we will use for computing the m-structure of $C \otimes_\ell \mathcal{F}(C)$ (this is the canonical acyclic twisted tensor product, defined in 2.3 on page 39). We will regard the Y-complexes as defining maps $C \to C^n \otimes \mathcal{F}(C)^n$, just as the coordinate coalgebra of C defines maps $C \to C^n$.

We define the following operator on $Y_{n,m}(\mathfrak{R}[C])$ almost exactly like the $\hat{\mathfrak{T}}$-map of $Z_{n,m}(\mathfrak{R}[C])$:

DEFINITION 1.2. Define $\tilde{\mathfrak{T}}\colon Y_{n,m}(\mathfrak{R}[C]) \to Y_{n,m}(\mathfrak{R}[C])$ by

$$\tilde{\mathfrak{T}}|\sigma^{-|\alpha|}\mathfrak{R}[C]_{|\alpha|+n}$$

$$= -q_{\infty,m} \circ \left\{ \sum_{j=1}^{|\alpha|} \sum_{i=1}^{\infty} (-1)^{j+|\alpha|i+ij} \downarrow^{|\alpha|+i-1} \circ (\Delta_i \circ_{j+n} *) \circ \uparrow^{|\alpha|} \right\}$$

where the $\{\Delta_k\}$ are part of the canonical $A(\infty)$-coalgebra structure of $\mathfrak{R}[C]$. The map $\tilde{\mathfrak{T}}\colon Y_{n,m}(\mathfrak{R}[C]) \to Y_{n,m}(\mathfrak{R}[C])$ will be of degree -1.

Remarks.   1.2.1. Compare this definition with equation (2.1) on page 42.

1.2.2. It is necessary to say something about precisely where the image of this version of $\tilde{\mathfrak{T}}$ lies. This is due to the fact that we have introduced some extra structure into $\sigma^{-|\alpha|}\mathfrak{R}[C]_{|\alpha|+n}$ — namely that of the sequence $\alpha = \{k_1, \ldots, k_n\}$. We will assume that the image of $\uparrow^{|\alpha|+i-1} \circ (\Delta_i \circ_j *) \circ \uparrow^{|\alpha|}$ lies in $\sigma^{-|\alpha(i,j)|}\mathfrak{R}[C]_{|\alpha(i,j)|+n}$, where $\alpha(i,j)$ is the sequence $\{k_1, \ldots, k_{r(j)} + i - 1, \ldots, k_n\}$, and $r(j)$ is the smallest value of t such that $j < \sum_{\ell=1}^{t} \alpha_\ell$.

PROPOSITION 1.3. *For any sequence of n nonnegative integers, α, let $\tilde{\mathfrak{Z}}$ be defined on $Y_n(\mathfrak{R}[C], \alpha)$ to be*

$$q_{\infty,m} \circ \left\{ \sum_{i=1}^{\infty} \sum_{t=1}^{n} (-1)^{|\alpha|i} \downarrow^{|\alpha|+i-1} \right.$$
$$\left. \circ \mathfrak{Z}\{(t, i-1), (n-t+\sum_{j=1}^{t-1} \alpha_j, \sum_{j=t}^{n} \alpha_j)\}(\Delta_i \circ_t *) \circ \uparrow^{|\alpha|} \right\}$$

Then the map, $\partial_Y = \tilde{\mathfrak{T}} + \tilde{\mathfrak{Z}} : Y_{n,m}(\mathfrak{R}[C]) \to Y_{n,m}(\mathfrak{R}[C])$, (where $\tilde{\mathfrak{T}}$ was defined in 1.2 above) has the property that $\partial_Y^2 = 0$. This means that $Y_{n,m}(\mathfrak{R}[C])$, equipped with this differential, constitutes a chain-complex.

Remarks.   1.3.1. Recall that $\mathcal{S}_n$ represents the *translation operator*, first defined in 2.17 on page 17. It translates the indices of a permutation up by n — so that $\mathcal{S}_n((1,2,3)) = (n+1, n+2, n+3)$.

1.3.2. Recall the *shuffle permutations*, $\mathfrak{Z}\{\beta_1, \ldots, \beta_k\}$, defined in 2.14 on page 45. No shuffling is performed by the term $\mathfrak{Z}\{(t, i-1), (n-t+\sum_{j=1}^{t-1} \alpha_j, \sum_{j=t}^{n} \alpha_j)\}$, when $i = 1$.

1.3.3. The differential of $Y_n(\mathfrak{R}[C], \alpha)$ was roughly designed to mimic the differential of the twisted tensor product $C \otimes_\ell \mathcal{F}(C)$, defined in 2.3 on page 39. Specifically, it is designed to mimic the differential of $(C \otimes_\ell \mathcal{F}(C))^n = C^n \otimes_L \mathcal{F}(C)^n$, where L is the twisting cochain $\downarrow \otimes \cdots \otimes \epsilon + \cdots + \epsilon \otimes \epsilon \otimes \cdots \otimes \downarrow$ (n factors), where $\epsilon : C \to \mathbb{Z} = C^0$ is the augmentation. The differential of $\mathcal{F}(C)^n$ is mimicked by $\hat{\mathfrak{T}}$, as in 2.25 on page 52 and the surrounding discussion. In 2.25, the maps $\mathcal{F}(\mathfrak{f}[C]_n)' : \mathcal{F}(C) \to \mathrm{Hom}_{\mathbb{Z}S_n}(\widehat{\mathcal{B}(Z_{n,*}(\mathfrak{R}[C]))}, \mathcal{F}(C)^n)$ were chain maps because the differential on $Z_{n,m}(\mathfrak{R}[C])$ mimicked the differential of $\mathcal{F}(C)$. We will prove the corresponding result for $Y_n(\mathfrak{R}[C], \alpha)$ and $C \otimes_\ell \mathcal{F}(C)$. In this case we must take the factor of C^n into account — this is responsible for the fact that the subscript of the composition-operation is $i + n - 1$ in this case, rather than $i - 1$.

The term $\mathfrak{Z}\{(t, i-1), (n-t+\sum_{j=1}^{t-1} \alpha_j, \sum_{j=t}^{n} \alpha_j)\}(\Delta_i \circ_t *) \circ \uparrow^{|\alpha|}$ represents the effect of the twisting cochain, ℓ, applied to the tth factor of C in C^n.

PROOF. We will relate $\tilde{\mathfrak{T}} + \tilde{\mathfrak{Z}}$ to the map $\hat{\mathfrak{T}}$, defined in 2.8 on page 42. Let $A(\alpha)_t =$

$\sum_{j=1}^{t-1}\alpha_j$. Then define

$$\tilde{\mathfrak{T}}(\alpha)_{t,i} = -\sum_{j=A(\alpha)_t}^{A(\alpha)_{t+1}}(-1)^{j+|\alpha|i+ij}\downarrow^{|\alpha|+i-1}\circ(\Delta_i\circ_{j+n}*)\circ\uparrow^{|\alpha|}$$

and

$$\tilde{\mathfrak{Z}}(\alpha)_{t,i} = (-1)^{|\alpha|i}\downarrow^{|\alpha|+i-1}$$

$$\circ\mathcal{Z}\{(t,i-1),(n-t+\sum_{j=1}^{t-1}\alpha_j,\sum_{j=t}^{n}\alpha_j)\}(\Delta_i\circ_t*)\circ\uparrow^{|\alpha|}$$

Then

$$\tilde{\mathfrak{T}} = \sum_\alpha\sum_{i=1}^{\infty}\sum_{t=1}^{n}\tilde{\mathfrak{T}}(\alpha)_{t,i}$$

and

$$\tilde{\mathfrak{Z}} = \sum_\alpha\sum_{i=1}^{\infty}\sum_{t=1}^{n}\tilde{\mathfrak{Z}}(\alpha)_{t,i}$$

We will actually prove that

$$\sum_{i+i'=v}\left(\tilde{\mathfrak{T}}(\alpha(i))_{t,i'}+\tilde{\mathfrak{Z}}(\alpha(i))_{t,i'}\right)\circ\left(\tilde{\mathfrak{T}}(\alpha)_{t,i}+\tilde{\mathfrak{Z}}(\alpha)_{t,i}\right) = 0$$

for all t, v, and sequences α, where $\alpha(i)$ represents the sequence that is equal to α in all terms, except for the tth, which has $i-1$ added to it.

This will imply the result because

(1.1) $$\tilde{\mathfrak{T}}_t\circ\tilde{\mathfrak{T}}_{t'} = -\tilde{\mathfrak{T}}_{t'}\circ\tilde{\mathfrak{T}}_t$$

(1.2) $$\tilde{\mathfrak{T}}_t\circ\tilde{\mathfrak{Z}}_{t'} = -\tilde{\mathfrak{Z}}_{t'}\circ\tilde{\mathfrak{T}}_t$$

(1.3) $$\tilde{\mathfrak{Z}}_t\circ\tilde{\mathfrak{Z}}_{t'} = -\tilde{\mathfrak{Z}}_{t'}\circ\tilde{\mathfrak{Z}}_t$$

where $t'\neq t$. This is due to C.1 on page 107.

Let $\gamma = \{\gamma_1,\ldots,\gamma_n\}$, where $\gamma_i = \alpha_i+1$. Now consider the isomorphism

$$\mathfrak{B}(\alpha)\colon Y_n(\mathfrak{R}[C],\alpha)\to\Sigma^{-|\gamma|}\mathfrak{R}_\gamma$$

(see 2.7.2 on page 42) defined by

$$\mathfrak{B}(\alpha) = \mathcal{Z}\{(1,\ldots,1),(\alpha_1,\ldots,\alpha_n)\}\circ\downarrow^n$$

We claim that

$$\tilde{\mathfrak{T}}(\alpha)_{t,i}+\tilde{\mathfrak{Z}}(\alpha)_{t,i} = \mathfrak{B}(\alpha(i))^{-1}\circ\hat{\mathfrak{T}}(\gamma)_{t,i}\circ\mathfrak{B}(\alpha)$$

where $\hat{\mathfrak{T}}(\gamma)_{t,i}$ is defined in equation (C.1) on page 107.

The statements in equations (1.1) through (1.3) follow from C.1 on page 107. The remaining statements (that the appropriate T-maps annihilate each other) follow from C.2 on page 108. $\square$

As in the discussion surrounding 2.23 on page 52, consider a canonical isomorphism
$$v(n,k)\colon\text{Hom}_{\mathbb{Z}S_{n+k}}(\mathfrak{R}[C]_{n+k},C^{n+k})\to\text{Hom}_{\mathbb{Z}S_{n+k}}(\Sigma^{-k}\mathfrak{R}[C]_{n+k},C^n\otimes(\Sigma^{-1}C)^k).$$

DEFINITION 1.4. Let n and k be integers ≥ 1, let C be a chain-complex, and let $\mathfrak{R}[C]$ be an f-resolution (defined in 3.1 on page 19) Define the map $v(n,k)$ to send $f \in \mathrm{Hom}_{\mathbb{Z}S_{n+k}}(\mathfrak{R}[C]_{n+k}, C^{n+k})$ to

$$\underbrace{1 \otimes \cdots \otimes 1}_{n \text{ factors}} \otimes \downarrow_k \circ f \circ \uparrow^k \in \mathrm{Hom}_{\mathbb{Z}S_{n+k}}(\Sigma^{-k}\mathfrak{R}[C]_{n+k}, C^n \otimes (\Sigma^{-1}C)^k)$$

Recall the notation $\downarrow_n$ — see 2.10 on page 43.

Now we can define the structure map of the twisted tensor product:

DEFINITION 1.5. Suppose $(C, \{\mathfrak{f}[C]_n : C \to \mathrm{Hom}_{\mathbb{Z}S_n}(\mathfrak{R}[C]_n, C^n)\})$ is a weakly-coherent m-coalgebra. For all $n \geq 0$, define the maps

$$\mathcal{A}(\mathfrak{f}[C]_n)' : C \otimes_\ell \mathcal{F}(C) \to \mathrm{Hom}_{\mathbb{Z}S_n}\left(Y_{n,m}(\mathfrak{R}[C]) \otimes \bar{B}(Z_{n,m}(\mathfrak{R}[C])), (C \otimes_\ell \mathcal{F}(C))^n\right)$$

by:

$$\mathcal{A}(\mathfrak{f}[C]_n)' = \sum_\alpha \mu(\alpha) \circ v(n, |\alpha|) \circ \mathfrak{f}[C]_{n+|\alpha|} \otimes \mathcal{F}(\mathfrak{f}[C]_n)' : C \otimes_\ell \mathcal{F}(C) \to$$

$$\mathrm{Hom}_{\mathbb{Z}S_n}\left(Y_{n,m}(\mathfrak{R}[C]) \otimes \bar{B}(Z_{n,m}(\mathfrak{R}[C])), (C \otimes_\ell \mathcal{F}(C))^n\right)$$

Here the target of the term $v(n, |\alpha|) \circ \mathfrak{f}[C]_{n+|\alpha|}$ lies in $C^n \otimes \subset C^n \otimes \mathcal{F}(C)^n$, and the summation is taken over all length-n sequences, α, of nonnegative integers. The map $\mu(\alpha) : C^n \otimes (\Sigma^{-1}C)^{\alpha_1} \otimes \cdots \otimes (\Sigma^{-1}C)^{\alpha_n} \otimes \mathcal{F}(C)^n \to (C \otimes_\ell \mathcal{F}(C))^n$ does the following:

1. performs a shuffle-isomorphism of the factors to get $(\Sigma^{-1}C)^{\alpha_1} \otimes \mathcal{F}(C) \otimes \cdots \otimes (\Sigma^{-1}C)^{\alpha_n} \otimes \mathcal{F}(C)$;
2. multiplies each factor of $\mathcal{F}(C)$ by its neighboring factor (after regarding that as being contained in $\mathcal{F}(C)$).

REMARK. 1.5.1. Note that, because the product-operation in $\mathcal{F}(C)$ is just the tensor product (i.e., it consists in re-labeling factors), statement 2 in the description of $\mu(\alpha)$ is not all that significant. From a computational point of view, $\mu(\alpha)$ is nothing but a *shuffling-operation*.

Now we will formalize the discussion given in remark 1.3.3 on page 70, regarding the differential of $Y_n(\mathfrak{R}[C], \alpha)$:

LEMMA 1.6. *Let $(C, \{\mathfrak{f}[C]_n : C \to \mathrm{Hom}_{\mathbb{Z}S_n}(\mathfrak{R}[C]_n, C^n)\})$ be a weakly coherent m-coalgebra. Let Y' denote the chain-complex whose underlying chain-modules are the same as $Y_{n,m}(\mathfrak{R}[C])$ as defined in 1.1 on page 69, but whose differential is defined in 1.2 on page 69. Let $\mathcal{A}(\mathfrak{f}[C]_n)'' = \sum_\alpha \mu(\alpha) \circ v(n, |\alpha|) \circ \mathfrak{f}[C]_{n+|\alpha|}$ be the restriction of the map*

$$\mathcal{A}(\mathfrak{f}[C]_n)' = \sum_\alpha \mu(\alpha) \circ v(n, |\alpha|) \circ \mathfrak{f}[C]_{n+|\alpha|} \otimes \mathcal{F}(\mathfrak{f}[C]_n)' : C \otimes_\ell \mathcal{F}(C) \to$$

$$\mathrm{Hom}_{\mathbb{Z}S_n}\left(Y_{n,m}(\mathfrak{R}[C]) \otimes \bar{B}(Z_{n,m}(\mathfrak{R}[C])), (C \otimes_\ell \mathcal{F}(C))^n\right)$$

(defined in 1.5 above), to $C = C \otimes 1 \subset C \otimes_\ell \mathcal{F}(C)$. Then $\mathcal{A}(\mathfrak{f}[C]_n)''$ is a chain-map:

$$\mathcal{A}(\mathfrak{f}[C]_n)'' \to \mathrm{Hom}_{\mathbb{Z}S_n}(Y_{n,m}(\mathfrak{R}[C]) \otimes \bar{B}(Z_{n,m}(\mathfrak{R}[C])), (C \otimes \mathcal{F}(C))^n)$$

Here C is equipped with its usual differential rather than the twisted differential of $C \otimes_\ell \mathcal{F}(C)$.

REMARK. 1.6.1. Note that the target of $\mathcal{A}(\mathfrak{f}[C]_n)''$ is
$\operatorname{Hom}_{\mathbb{Z}S_n}(Y_{n,m}(\mathfrak{R}[C])$ $\otimes$ $\bar{\mathcal{B}}(Z_{n,m}(\mathfrak{R}[C])), (C \otimes \mathcal{F}(C))^n)$ where
$C \otimes \mathcal{F}(C)$ is an *untwisted* tensor product.

PROOF. As before, we equip C with the canonical $A(\infty)$-coalgebra structure,
$\{\Delta_i : C \to C^i\}$, induced by its weakly-coherent m-structure — see 1.5 on page 36.

Recall the notation $\downarrow_i$ defined in 2.10 on page 43. This follows by an argument
very similar to the proof of 2.25 on page 52. Since the factors $\mu(\alpha)$ in the expression
for $\mathcal{A}(\mathfrak{f}[C]_n)'$ simply shuffle the factors of C and $\mathcal{F}(C)$, it suffices to prove that the
map $Z = \sum_\alpha v(n, |\alpha|) \circ \mathfrak{f}[C]_{n+|\alpha|} : C \to \operatorname{Hom}_{\mathbb{Z}S_n}(Y_{n,m}(\mathfrak{R}[C]), C^n \otimes \mathcal{F}(C)^n)$ is a
chain-map.

The boundary of $C \otimes_\ell \mathcal{F}(C)$ is given by 2.1 on page 38 — it is

$$\partial_C \otimes 1 - 1 \otimes \sum_{i=1}^{\infty} \underbrace{1 \otimes \cdots \otimes \downarrow_i \circ \Delta_i \circ \uparrow \otimes \cdots \otimes 1}_{j^{\text{th}} \text{ position}}$$

The fact that $(C, \{\mathfrak{f}[C]_n : C \to \operatorname{Hom}_{\mathbb{Z}S_n}(\mathfrak{R}[C]_n, C^n)\})$ is weakly coherent implies the
result. The weak coherence implies that the term, $\underbrace{1_C \otimes 1 \otimes \cdots \otimes \downarrow_i \circ \Delta_i \circ \uparrow \otimes \cdots \otimes 1}_{j^{\text{th}} \text{ position}}$,

evaluated on the image of $\{\mathfrak{f}[C]_n\}$ in $\operatorname{Hom}_{\mathbb{Z}S_n}(*, *)$, produces the same result as the
formal composition-operation in $(-1)^{j+|\alpha|i+ij} \downarrow^{|\alpha|+i-1} \circ (\Delta_i \circ_{j+n} *) \circ \uparrow^{|\alpha|}$ in the def-
inition of $\widetilde{\mathfrak{T}} : Y_{n,m}(\mathfrak{R}[C]) \to Y_{n,m}(\mathfrak{R}[C])$ in 1.2 on page 69. The sign of $(-1)^{j+|\alpha|i+ij}$
results from the Koszul convention. Now we present the details:

The composition

$$1_C \otimes \underbrace{1 \otimes \cdots \otimes \downarrow_i \circ \Delta_i \circ \uparrow \otimes \cdots \otimes 1}_{j^{\text{th}} \text{ position}} \circ v(n, |\alpha|) \circ \mathfrak{f}[C]_{n+|\alpha|}$$

$$= 1^n \otimes \underbrace{1 \otimes \cdots \otimes \downarrow_i \circ \Delta_i \circ \uparrow \otimes \cdots \otimes 1}_{j^{\text{th}} \text{ position}} \circ 1^n \otimes \downarrow_{|\alpha|} \circ \mathfrak{f}[C]_{n+|\alpha|} \circ \uparrow^{|\alpha|}$$

(see the definition of $v(n, |\alpha|)$ in 1.4 on page 72).

Recall the notation $\downarrow_i$ defined in 2.10 on page 43. We eliminate the superfluous
factors of 1^n and slide $\downarrow_i \circ \Delta_i \circ \uparrow$ past $j-1$ copies of $\downarrow$ to get a sign of $(-1)^{j-1}$ (since
$\downarrow$ and $\downarrow_i \circ \Delta_i \circ \uparrow$ are both of degree -1). The factor of $\uparrow$ cancels a corresponding factor
of $\downarrow$. Now we slide the Δ_i past the remaining $|\alpha| - j$ copies of $\downarrow$ to get another factor
of $(-1)^{i(|\alpha|-j)}$. This is due to the fact that Δ_i is of degree $i - 2 \equiv i \pmod{2}$, and $\downarrow$ is
of degree -1. To summarize so far, we have:

$$1^n \otimes \underbrace{1 \otimes \cdots \otimes \downarrow_i \circ \Delta_i \circ \uparrow \otimes \cdots \otimes 1}_{j^{\text{th}} \text{ position}} \circ 1^n \otimes \downarrow_{|\alpha|} \circ \mathfrak{f}[C]_{n+|\alpha|} \circ \uparrow^{|\alpha|}$$

$$= -(-1)^{j+|\alpha|i+ij} \downarrow_{|\alpha|+i-1} \circ \underbrace{1 \otimes \cdots \otimes \Delta_i \otimes \cdots \otimes 1}_{j^{\text{th}} \text{ position}} \circ \mathfrak{f}[C]_{n+|\alpha|} \circ \uparrow^{|\alpha|}$$

Statement 2 (on page 19) in the definition of a weakly-coherent m-coalgebra implies that, for all $c \in C$:

$$\underbrace{1 \otimes \cdots \otimes \Delta_i \otimes \cdots \otimes 1}_{j^{\text{th}} \text{ position}} = (-1)^{\dim(c)i} \mathfrak{f}[C]_{n+|\alpha|+i-1}(c) \circ (\Delta_i \circ_j *)$$

The boundary of an element of $z \in \operatorname{Hom}_{\mathbb{Z} S_n}(Y_{n,m}(\mathfrak{R}[C]), C^n \otimes \mathcal{F}(C)^n)$ is induced by its boundary in $\operatorname{Hom}_{\mathbb{Z} S_n}(Y_{n,m}(\mathfrak{R}[C]), C^n \otimes \mathcal{F}(C)^n)$, which is $\partial_{\mathcal{F}} \circ z - (-1)^{\dim(z)} z \circ \partial_Y$. In the present situation

$$\partial_{\mathcal{F}} \circ z(c) = -(-1)^{j+|\alpha|i+ij} \downarrow_{|\alpha|+i-1} \circ \underbrace{1 \otimes \cdots \otimes \Delta_i \otimes \cdots \otimes 1}_{j^{\text{th}} \text{ position}}$$

$$\circ \mathfrak{f}[C]_{n+|\alpha|} \circ \uparrow^{|\alpha|}(c)$$

(1.4)

$$= -(-1)^{j+|\alpha|i+ij}(-1)^{\dim(c)}(-1)^{\dim(c)|\alpha|} \downarrow_{|\alpha|+i-1} \circ \mathfrak{f}[C]_{n+|\alpha|}(c) \circ (\Delta_i \circ_j *) \circ \uparrow^{|\alpha|}$$

(Recall that $\downarrow_{|\alpha|+i-1} = \underbrace{\downarrow \otimes \cdots \otimes \downarrow}_{|\alpha|+i-1 \text{ copies}}$ — see 2.10 on page 43). (the extra factor of

$(-1)^{\dim(c)|\alpha|}$ resulted from slipping c past the term of $\uparrow^{|\alpha|}$); and

$$(-1)^{\dim(c)} z \circ \partial_Y v(n, |\alpha|) \circ \mathfrak{f}[C]_{n+|\alpha|}(c)$$

$$= -(-1)^{j+|\alpha|i+ij} v(n, |\alpha|) \circ \mathfrak{f}[C]_{n+|\alpha|}(c) \circ \downarrow^{|\alpha|+i-1} \circ (\Delta_i \circ j*) \circ \uparrow^{|\alpha|}$$

$$= -(-1)^{\dim(c)(|\alpha|+i)}(-1)^{j+|\alpha|i+ij} \downarrow^{|\alpha|+i-1} \circ v(n, |\alpha|)$$

$$\circ \mathfrak{f}[C]_{n+|\alpha|}(c) \circ (\Delta_i \circ j*) \circ \uparrow^{|\alpha|}$$

since we have slipped $\downarrow^{|\alpha|+i-1}$ past $v(n, |\alpha|) \circ \mathfrak{f}[C]_{n+|\alpha|}(c)$. The difference of these two terms (equation (1.4) and equation (1.4)) is clearly zero.

This implies the conclusion, since the resulting boundary of $\operatorname{Hom}_{\mathbb{Z} S_n}\left(Y_{n,m}(\mathfrak{R}[C]) \otimes \bar{B}(Z_{n,m}(\mathfrak{R}[C])), (C \otimes_\ell \mathcal{F}(C))^n\right)$ is the same as that of $(s(|\alpha|))^{-1} \operatorname{Hom}_{\mathbb{Z} S_n}(\mathfrak{R}[C]_n, C^n)$. $\square$

LEMMA 1.7. *Let* $(C, \{\mathfrak{f}[C]_n : C \to \operatorname{Hom}_{\mathbb{Z} S_n}(\mathfrak{R}[C]_n, C^n)\})$ *be a weakly coherent m-coalgebra. Let* $Y_{n,m}(\mathfrak{R}[C])$ *be exactly as defined in 1.1 on page 69 with a differential of* $\tilde{\mathfrak{T}} + \tilde{3}$ *(as defined in 1.2 on page 69). Consider the map*

$$\mathcal{A}(\mathfrak{f}[C]_n)' = \sum_\alpha \mu(\alpha) \circ v(n, |\alpha|) \circ \mathfrak{f}[C]_{n+|\alpha|} \otimes \mathcal{F}(\mathfrak{f}[C]_n)' : C \otimes_\ell \mathcal{F}(C)$$

$$\to \operatorname{Hom}_{\mathbb{Z} S_n}\left(Y_{n,m}(\mathfrak{R}[C]), (C \otimes_\ell \mathcal{F}(C))^n\right)$$

(defined in 1.5 on page 72), restricted to $C = C \otimes 1 \subset C \otimes_\ell \mathcal{F}(C)$. *Then* $\mathcal{A}(\mathfrak{f}[C]_n)'$ *is a chain-map. Here,* $C \otimes 1$ *is assumed to have the untwisted differential.*

PROOF. As before, the fact that the factors $\mu(\alpha)$ in the expression for $\mathcal{A}(\mathfrak{f}[C]_n)'$ simply shuffle the factors of C and $\mathcal{F}(C)$ implies that it suffices to prove that the map $Z = \sum_\alpha v(n, |\alpha|) \circ \mathfrak{f}[C]_{n+|\alpha|} : C \to \operatorname{Hom}_{\mathbb{Z} S_n}(Y_{n,m}(\mathfrak{R}[C]), C^n \otimes_L \mathcal{F}(C)^n))$ is a chain-map. Here $L : C^n \to \mathcal{F}(C)^n$ is the twisting cochain $\downarrow \otimes \cdots \otimes \epsilon + \epsilon \otimes \downarrow \otimes \cdots \otimes \epsilon + \cdots + \epsilon \otimes \cdots \otimes \downarrow$, where $\epsilon : C \to \mathbb{Z} \subset \mathcal{F}(C)_0$ is the augmentation. There is a canonical isomorphism of twisted tensor products $C^n \otimes_L \mathcal{F}(C)^n \cong (C \otimes_\ell \mathcal{F}(C))^n$. The only elements in the present lemma that distinguish it from 1.6 above are: the twisting cochain, L; and[1]

$$\tilde{\mathfrak{z}} = \sum_\alpha \sum_{i=2}^\infty \sum_{t=2}^n (-1)^{|\alpha|i} \downarrow^{|\alpha|+i-1}$$

$$\circ \, \mathfrak{Z}\{(t, i-1), (n-t+\sum_{j=1}^{t-1}\alpha_j, \sum_{j=t}^n \alpha_j)\}(\Delta_i \circ_t *) \circ \uparrow^{|\alpha|}$$

The fact that $(C, \{\mathfrak{f}[C]_n : C \to \text{Hom}_{\mathbb{Z}S_n}(\mathfrak{R}[C]_n, C^n)\})$ is weakly coherent implies the result. The effect of the twisting cochain on the tth factor is to:

1. evaluate Δ_i on it;
2. shuffle all but one of the resulting new factors of C (there will be i of them — they are the result of applying Δ_i) to the tth copy of $\mathcal{F}(C)$.

Let $c \in C$ be some arbitrary element. The weak-coherency of $(C, \{\mathfrak{f}[C]_n : C \to \text{Hom}_{\mathbb{Z}S_n}(\mathfrak{R}[C]_n, C^n)\})$ and the Koszul convention imply that the result of statement 1 on

$$v(n, |\alpha|) \circ \left(\mathfrak{f}[C]_{n+|\alpha|} \circ \uparrow^{|\alpha|}\right)(c) = (-1)^{\dim(c)|\alpha|}\mathfrak{f}[C]_{n+|\alpha|}(c) \circ \uparrow^{|\alpha|}$$

is

$1'$.

$$(-1)^{\dim(c)|\alpha|} \underbrace{1 \otimes \cdots \otimes \downarrow_i \circ \Delta_i \otimes \cdots \otimes 1}_{t^{\text{th}} \text{ position}} \circ 1^n \otimes \downarrow_{|\alpha|} \mathfrak{f}[C]_{n+|\alpha|}(c) \circ \uparrow^{|\alpha|}$$

where $t \le n$

$$= (-1)^{\dim(c)|\alpha|}(-1)^{|\alpha|i} \underbrace{1 \otimes \cdots \otimes \downarrow_i \otimes \cdots \otimes 1}_{t^{\text{th}} \text{ position}} \otimes 1$$

$$\otimes \underbrace{1 \otimes \cdots \otimes \Delta_i \otimes \cdots \otimes 1}_{t^{\text{th}} \text{ position}} \circ \mathfrak{f}[C]_{n+|\alpha|}(c) \uparrow^{|\alpha|}$$

$$= (-1)^{\dim(c)(i+|\alpha|)}(-1)^{|\alpha|i} 1 \otimes \cdots \otimes \downarrow_i \otimes \cdots \otimes 1 \otimes \mathfrak{f}[C]_{n+|\alpha|}(\Delta_i \circ_t *) \circ \uparrow^{|\alpha|}$$

(by 1.6 on page 37). Here $\downarrow_i = \underbrace{\downarrow \otimes \cdots \otimes \downarrow}_{i \text{ copies}}$ — see 2.10 on page 43.

The result of 2 on this is given by:

$2'$. the effect of the term $\mathfrak{Z}\{(t, i-1), (n-t+\sum_{j=1}^{t-1}\alpha_j, \sum_{j=t}^n \alpha_j)\}$. The permutation $\mathfrak{Z}\{(t, i-1), (n-t+\sum_{j=1}^{t-1}\alpha_j, \sum_{j=t}^n \alpha_j)\}$ shuffles the sequences $(t, i-1)$ and $(n-t+\sum_{j=1}^{t-1}\alpha_j, \sum_{j=t}^n \alpha_j)$, where each integer in each sequences represents a block of objects to be shuffled together — it is a permutation that is composed with Δ_i.

---

[1] Recall that C has the structure of an $A(\infty)$-coalgebra, $\{\Delta_i : C \to C^i\}$, induced by its weakly-coherent m-structure — see 1.5 on page 36.

As in 1.6 on page 72, we compute the boundary of an element of $z \in \operatorname{Hom}_{\mathbb{Z}S_n}(Y_{n,m}(\mathfrak{R}[C]), C^n \otimes \mathcal{F}(C)^n)$ as $\partial_{\mathcal{F}} \circ z - (-1)^{\dim(z)} z \circ \partial_Y$. In the present situation, $\partial_{\mathcal{F}} \circ z(c) = P(-1)^{\dim(c)(i+|\alpha|)}(-1)^{|\alpha|i} 1 \otimes \cdots \otimes \underbrace{1 \otimes \cdots \otimes \downarrow_i \otimes \cdots \otimes 1}_{t^{\text{th}} \text{ position}} \otimes \cdots \otimes 1 \otimes \mathfrak{f}[C]_{n+|\alpha|}(c)(\Delta_i \circ_t *) \circ \uparrow^{|\alpha|}$, where

$$
P = \mathcal{Z}\big\{ (t, i-1), (n - t + \sum_{j=1}^{t-1} \alpha_j, \sum_{j=t}^{n} \alpha_j) \big\}
$$

In the present context

$$
z \circ \partial_Y = (-1)^{|\alpha|i} v(n, |\alpha|) \circ \mathfrak{f}[C]_{n+|\alpha|}(c) \downarrow^{|\alpha|+i-1} \circ P(\Delta_i \circ t*) \circ \uparrow^{|\alpha|}
$$

$$
= (-1)^{|\alpha|i} \circ \downarrow^{|\alpha|}(\mathfrak{f}[C]_{n+|\alpha|}) \circ \uparrow^{|\alpha|}(c) \downarrow^{|\alpha|+i-1} \circ P(\Delta_i \circ t*) \circ \uparrow^{|\alpha|}
$$

$$
= (-1)^{\dim(c)|\alpha|}(-1)^{|\alpha|i} \circ \downarrow^{|\alpha|}(\mathfrak{f}[C]_{n+|\alpha|}(c)) \circ \uparrow^{|\alpha|} \downarrow^{|\alpha|+i-1} \circ \\
P(\Delta_i \circ t*) \circ \uparrow^{|\alpha|}
$$

$$
= (-1)^{\dim(c)|\alpha|}(-1)^{|\alpha|i} \circ \downarrow^{|\alpha|}(\mathfrak{f}[C]_{n+|\alpha|}(c)) \circ \downarrow^{i-1} \circ P(\Delta_i \circ t*) \circ \uparrow^{|\alpha|}
$$

$$
= (-1)^{\dim(c)|\alpha|}(-1)^{|\alpha|i} \circ P \downarrow^{|\alpha|}(\mathfrak{f}[C]_{n+|\alpha|}(c)) \circ \downarrow^{i-1}(\Delta_i \circ t*) \circ \uparrow^{|\alpha|}
$$

$$
= (-1)^{\dim(c)(|\alpha|+i-1)}(-1)^{|\alpha|i} \circ P \\
\downarrow^{|\alpha|+i-1} v(n, |\alpha|) \circ (\mathfrak{f}[C]_{n+|\alpha|}(c)) \circ (\Delta_i \circ t*) \circ \uparrow^{|\alpha|}
$$

And, if we form the difference, $\partial_{\mathcal{F}} \circ z - (-1)^{\dim(z)} z \circ \partial_Y$, we get zero.

The upshot of all of this is the fact that weak coherence guarantees that the effect of $\tilde{3}$ on $Y_{n,m}(\mathfrak{R}[C])$ coincides with the effect of the twisting cochain, L, on $C^n \otimes_L \mathcal{F}(C)^n$. The contribution of both terms to the boundary in $\operatorname{Hom}_{\mathbb{Z}S_n}(Y_{n,m}(\mathfrak{R}[C]), (C \otimes_\ell \mathcal{F}(C))^n)$ cancels out, and the de-suspended structure maps $\mathcal{A}(\mathfrak{f}[C]_n)'|C \otimes 1 = \sum_\alpha \mu(\alpha) \circ v(n, |\alpha|) \circ \mathfrak{f}[C]_{n+|\alpha|}: C \to \operatorname{Hom}_{\mathbb{Z}S_n}(Y_{n,m}(\mathfrak{R}[C]), (C \otimes_\ell \mathcal{F}(C))^n)$ remain chain-maps.

$\square$

Recall the k-linear product $t_k: \mathfrak{R}_{i_1} \otimes \cdots \otimes \mathfrak{R}_{i_k} \to \mathfrak{R}_s$ of degree $k - 2$, defined in remark 2.14.2 on page 45. It was defined by:

$$
t_k(r_1 \otimes \cdots \otimes r_k) = (-1)^{\sum_{i=1}^{k} \dim(r_i)} r_1 \circ_1 \cdots r_k \circ_k \Delta_k
$$

The basic property of the t_k is expressed by remark 2.14.2 on page 45 and, particularly, the diagram in that remark.

Now we find a differential on $Y_{n,m}(\mathfrak{R}[C]) \otimes \bar{\mathcal{B}}(Z_{n,m}(\mathfrak{R}[C]))$ that makes the map $\mathcal{A}(\mathfrak{f}[C]_n)'$ a chain-map. To this end we compute an action of $\bar{\mathcal{B}}(Z_{n,m}(\mathfrak{R}[C]))$ on $Y_{n,m}(\mathfrak{R}[C])$.

DEFINITION 1.8. Let $\beta_1, \ldots, \beta_k$ be k sequences of length n of nonnegative integers, and let α be the elementwise sum of these sequences. We define a natural k-linear pairing $\hat{a}_k \colon Y_{n,m}(\mathfrak{R}[C]) \otimes Z_{n,m}(\mathfrak{R}[C]) \otimes \cdots \otimes Z_{n,m}(\mathfrak{R}[C]) \to Y_{n,m}(\mathfrak{R}[C])$ via

$$\hat{a}_k = (-1)^{k|\alpha|} \downarrow^{|\alpha|} \mathcal{S}_n(\mathcal{Z}\{\beta_1, \ldots, \beta_k\}) \circ t_k \circ \uparrow^{|\beta_1|} \otimes \cdots \otimes \uparrow^{|\beta_k|}$$

$$\colon \Sigma^{-|\beta_1|} \mathfrak{R}[C]_{n+|\beta_1|} \otimes \cdots \otimes \Sigma^{-|\beta_k|} \mathfrak{R}[C]_{n+|\beta_k|} \to \Sigma^{-|\alpha|} \mathfrak{R}[C]_{|\alpha|+n}$$

where $\alpha = \sum_{i=1}^{k} \beta_i$ and the summation of sequences is taken elementwise. Recall that $\mathcal{S}_n \colon S_k \to S_{k+n}$ is the map that shifts all indices up by n.

REMARK. 1.8.1. Note that this is almost exactly the same as the definition of the $\{\bar{\mu}_k\}$ in 2.15 on page 45 — we only make a distinction because the product here is regarded as a pairing $Y_{n,m}(\mathfrak{R}[C]) \otimes Z_{n,m}(\mathfrak{R}[C]) \otimes \cdots \otimes Z_{n,m}(\mathfrak{R}[C]) \to Y_{n,m}(\mathfrak{R}[C])$ that defines $Y_{n,m}(\mathfrak{R}[C])$ to be a right $A(\infty)$-module over $Z_{n,m}(\mathfrak{R}[C])$.

Now we are in a position to define the boundary-operator of $Y_{n,m}(\mathfrak{R}[C]) \otimes \bar{\mathcal{B}}(Z_{n,m}(\mathfrak{R}[C]))$:

THEOREM 1.9. *The map*

$$\partial_{Y \otimes \mathcal{F}} = \partial_Y \otimes 1 + 1 \otimes \partial_{\mathcal{F}} - \sum_{i=2}^{\infty} \hat{\mu}_i \circ 1 \otimes \underbrace{\downarrow \otimes \cdots \otimes \downarrow}_{i-1 \, factors} \colon Y_{n,m}(\mathfrak{R}[C]) \otimes \bar{\mathcal{B}}(Z_{n,m}(\mathfrak{R}[C]))$$

$$\to Y_{n,m}(\mathfrak{R}[C]) \otimes \bar{\mathcal{B}}(Z_{n,m}(\mathfrak{R}[C]))$$

is a differential, where

$$\hat{a}_k \circ 1 \otimes \underbrace{\downarrow \otimes \cdots \otimes \downarrow}_{k-1 \, factors} (y \otimes \Sigma r_1 \otimes \cdots \otimes \Sigma r_j)$$

$$= \hat{a}_k \circ 1 \otimes \underbrace{\downarrow \otimes \cdots \otimes \downarrow}_{k-1 \, factors} (y \otimes \Sigma r_1 \otimes \cdots \otimes \Sigma r_{k-1}) \otimes \Sigma r_k \otimes \cdots \otimes \Sigma r_j$$

if $j \geq k - 1$, and is defined to vanish if $j < k - 1$, and $y \in Y_{n,m}(\mathfrak{R}[C], \beta_1)$, $r_i \in Z_{n,m}(\mathfrak{R}[C], \beta_{i+1})$, $\{\beta_1, \ldots, \beta_j\}$ are sequences of length n of nonnegative integers, and $[r_1| \ldots |r_j]$ is identified with $\Sigma r_1 \otimes \cdots \otimes \Sigma r_j$. Furthermore the structure map

$$\mathcal{A}(\mathfrak{f}[C]_n)' = \sum_{\alpha} \mu(\alpha) \circ v(n, |\alpha|) \circ \mathfrak{f}[C]_{n+|\alpha|} \otimes \mathcal{F}(\mathfrak{f}[C]_n)' \colon C \otimes_{\ell} \mathcal{F}(C)$$

$$\to \mathrm{Hom}_{\mathbb{Z} S_n}\left(\{Y_{n,m}(\mathfrak{R}[C]) \otimes \bar{\mathcal{B}}(Z_{n,m}(\mathfrak{R}[C])), \partial_{Y \otimes \mathcal{F}}\}, (C \otimes_{\ell} \mathcal{F}(C))^n\right)$$

is a chain map.

Remarks. 1.9.1. Recall that $C \otimes_{\ell} \mathcal{F}(C)$ is the canonical acyclic twisted tensor product, defined in 2.3 on page 39.

1.9.2. $Y_{n,m}(\mathfrak{R}[C]) \otimes \mathcal{F}(\mathfrak{R}[C], n)$ is nothing but a *twisted tensor product* (as defined in 2.5 on page 40) with base $\mathcal{F}(\mathfrak{R}[C], n)$ and fiber $Y_{n,m}(\mathfrak{R}[C])$. Here the twisting cochain is $\ell \colon \mathcal{F}(\mathfrak{R}[C], n) \to Z_{n,m}(\mathfrak{R}[C])$, and this is defined to vanish on any elements $[r_1| \ldots |r_k]$ with $k \neq 1$ (by abuse of notation, we are denoting some elements of $\mathcal{F}(\mathfrak{R}[C], n)$ as elements of the formal bar construction $\bar{\mathcal{B}}(Z_{n,m}(\mathfrak{R}[C]))$ — see 2.4 on page 40). In addition, as per remark 1.8.1 above, we must regard $Y_{n,m}(\mathfrak{R}[C])$ as an

$A(\infty)$-module over $Z_{n,m}(\mathfrak{R}[C])$ and we must reverse the directionality of all actions (so it is a *right-module* instead of a *left* module, etc).

PROOF. Suppose $\{\Delta_i\colon C \to C^i\}$ stands for the canonical $A(\infty)$-coalgebra structure induced by its weakly-coherent m-structure — see 1.5 on page 36. The proof of the theorem is in several steps:

1. Proof that $\partial_{Y \otimes \mathcal{F}}$ is a differential:
    a. Claim: $\partial_Y \circ = (-1)^i \hat{\mu}_i \circ \partial_Y \otimes 1$. This is due to the fact that, after all of the suspensions and desuspensions are done both expressions are sums of terms of the form $\pm \Delta_i \circ_t r_1 \circ_1 \ldots r_k \circ_k \Delta_m$, where the original expression was evaluated on $r_1 \otimes \cdots \otimes r_k$. The $\{\Delta_i\}$ on the *left* come from ∂_Y and the $\{\Delta_m\}$ on the right come from $\hat{\mu}_i$ — and these two operations never *directly interact* with each other. The changes in sign due to suspensions and desuspensions will also be the same in both cases. The only difference in the two expressions will occur in the transition from to *left-composites* with Δ_i — see 1.8 above, and remark 2.14.2 on page 45. In $\hat{\mu}_i \circ \partial_Y \otimes 1$ the terms will be of one dimension lower when converted into composites with Δ_i so that corresponding signs that go with this transition (see remark 2.14.2) will be altered by a factor of $(-1)^i$.

    This claim and the fact that $\hat{\mu}_i \circ 1 \otimes \ell \otimes \cdots \otimes \ell \circ \partial_Y \otimes 1 = (-1)^{i-1} \circ \partial_Y \otimes 1 \circ 1 \otimes \ell \otimes \cdots \otimes \ell$, imply that

    $$\partial_Y \otimes 1 \circ \sum_{i=2}^{\infty} \hat{\mu}_i \circ \underbrace{\ell \otimes \cdots \otimes \ell}_{i-1 \text{ factors}} + \sum_{i=2}^{\infty} \hat{\mu}_i \circ \underbrace{\ell \otimes \cdots \otimes \ell}_{i-1 \text{ factors}} \circ \partial_Y \otimes 1 = 0$$

    b. It remains to prove that

    $$\left(1 \otimes \partial_{\mathcal{F}} - \sum_{i=2}^{\infty} \hat{\mu}_i \circ \underbrace{\ell \otimes \cdots \otimes \ell}_{i-1 \text{ factors}} \right)^2 = 0$$

    This *second* fact follows from the fact that

    $$1 \otimes \partial_{\mathcal{F}} - \sum_{i=2}^{\infty} \hat{\mu}_i \circ \underbrace{\ell \otimes \cdots \otimes \ell}_{i-1 \text{ factors}}$$

    is essentially the *same* as the boundary map of $\mathcal{F}(\mathfrak{R}[C])$ (see 2.14 on page 45, 1.7 on page 74, and 2.11 on page 43) — it only differs slightly in its treatment of the first factor.

2. Proof that $\mathcal{A}(\mathfrak{f}[C]_n)'$ is a chain-map: First, note that

$$\mathcal{A}(\mathfrak{f}[C]_n)' = \sum_{\alpha} \mu(\alpha) \circ v(n, |\alpha|) \circ \mathfrak{f}[C]_{n+|\alpha|} \otimes \mathcal{F}(\mathfrak{f}[C]_n)' \colon C \otimes \mathcal{F}(C)$$

$$\to \operatorname{Hom}_{\mathbb{Z}S_n}\left(Y_{n,m}(\mathfrak{R}[C]) \otimes \bar{\mathcal{B}}(Z_{n,m}(\mathfrak{R}[C])), (C \otimes_\ell \mathcal{F}(C))^n \right)$$

is a chain-map, where $C \otimes \mathcal{F}(C)$ and $Y_{n,m}(\mathfrak{R}[C]) \otimes \bar{\mathcal{B}}(Z_{n,m}(\mathfrak{R}[C]))$ are ordinary (i.e., untwisted) tensor products. This claim follows from 1.7 on page 74 and the fact that $\mathcal{F}(\mathfrak{f}[C]_n)'\colon \mathcal{F}(C) \to \operatorname{Hom}_{\mathbb{Z}S_n}(\bar{\mathcal{B}}(Z_{n,m}(\mathfrak{R}[C])), \mathcal{F}(C)^n)$

is a chain-map, as was proved in theorem 2.25 on page 52. The map $\mathcal{A}(\mathfrak{f}[C]_n)'$ in this claim, is the result of forming the tensor product of $\mathcal{A}(\mathfrak{f}[C]_n)'|C \otimes 1$ and $\mathcal{F}(\mathfrak{f}[C]_n)'$.

We must, consequently, prove that the following two perturbations of the differentials of $C \otimes \mathcal{F}(C)$ and $Y_{n,m}(\mathfrak{R}[C]) \otimes \bar{\mathcal{B}}(Z_{n,m}(\mathfrak{R}[C]))$ are compatible:

1. the twisting term of the twisting cochain $\ell \colon C \to \mathcal{F}(C)$ — this is

$$\sum_{i=2}^{\infty} (1 \otimes \underbrace{\ell \otimes \cdots \otimes \ell}_{i-1 \text{ factors}} \otimes 1) \circ (\Delta_i \otimes 1) \colon C \otimes \mathcal{F}(C) \to C \otimes \mathcal{F}(C)$$

and;

2. $-\sum_{i=2}^{\infty} \hat{\mu}_i \circ \underbrace{\ell \otimes \cdots \otimes \ell}_{i-1 \text{ factors}} \colon Y_{n,m}(\mathfrak{R}[C]) \otimes \bar{\mathcal{B}}(Z_{n,m}(\mathfrak{R}[C])) \to Y_{n,m}(\mathfrak{R}[C]) \otimes$

$\bar{\mathcal{B}}(Z_{n,m}(\mathfrak{R}[C]))$. This follows from the diagram in remark 2.14.2 on page 45.

$\square$

REMARK. 1.9.1. Theorem 1.9 on page 77 essentially equips the tensor product $Y_{n,m}(\mathfrak{R}[C]) \otimes \bar{\mathcal{B}}(Z_{n,m}(\mathfrak{R}[C]))$ with a twisted differential that makes it the twisted tensor product $Y_{n,m}(\mathfrak{R}[C]) \bigstar_{\ell,\{\hat{\mu}_i\}} \bar{\mathcal{B}}(Z_{n,m}(\mathfrak{R}[C]))$.

PROPOSITION 1.10. *The chain-complexes* $\{Y_{n,m}(\mathfrak{R}[C]) \bigstar_{\ell,\{\hat{\mu}_i\}} \bar{\mathcal{B}}(Z_{n,m}(\mathfrak{R}[C]))\}$ *are acyclic in positive dimensions.*

Remarks. 1.10.1. This follows by repeated spectral-sequence arguments, proving that $Y_{n,m}(\mathfrak{R}[C])$ is acyclic (first with the differential $\tilde{\mathfrak{T}}$, then with $\tilde{\mathfrak{T}} + \tilde{\mathfrak{Z}}$), and finally proving $\{Y_{n,m}(\mathfrak{R}[C]) \bigstar_{\ell,\{\hat{\mu}_i\}} \bar{\mathcal{B}}(Z_{n,m}(\mathfrak{R}[C]))\}$ is acyclic with the full differential given above.

We are now in a position to complete our construction of the coordinate coalgebra of $C \otimes_{\ell} \mathcal{F}(C)$. The final step in the construction is motivated by the desire to define natural morphisms of m-coalgebras:

1. $p \colon C \otimes_{\ell} \mathcal{F}(C) \to C$;
2. $\iota \colon \mathcal{F}(C) \to C \otimes_{\ell} \mathcal{F}(C)$;

Map 1 is just the projection to the base, and map 2 is the inclusion of the fiber. If $\{A_n\}$ is the coordinate coalgebra of $C \otimes_{\ell} \mathcal{F}(C)$, then the existence of map 1 requires the existence of a morphism of f-resolutions: $\{\mathfrak{R}[C]_n \to A_n\}$, and map 2 requires the existence of $\{A_n \to \mathcal{F}(\mathfrak{R}[C], n)\}$. We will also require that $C \otimes_{\ell} \mathcal{F}(C)$ be a right strict m-module over $\mathcal{F}(C)$. Recall the definition of an m-module in 3.14 on page 27.

This requirement essentially imposes the additional requirement that we form the inverse limit of the inverse system

$$Y_{n,m'}(\mathfrak{R}[C])\bigstar_{\ell,\{\hat{\mu}_i\}}\bar{\mathcal{B}}(Z_{n,m''}(\mathfrak{R}[C]))$$

$$\Big\downarrow {\scriptstyle q_{m',m}\otimes p_{m'',m}}$$

$$Y_{n,m}(\mathfrak{R}[C])\bigstar_{\ell,\{\hat{\mu}_i\}}\bar{\mathcal{B}}(Z_{n,m}(\mathfrak{R}[C]))$$

since:

1. We want to construct a coordinate coalgebra of $C\otimes_\ell \mathcal{F}(C)$, and we want $C\otimes_\ell \mathcal{F}(C)$ to be an m-module over $\mathcal{F}(C)$;
2. $\mathcal{F}(\mathfrak{R}[C],n)$ is a form of completion of $\bar{\mathcal{B}}(Z_{n,m}(\mathfrak{R}[C]))$ — see 2.28 on page 56.

1.10.2. The discussion of $Y_{n,m}(\mathfrak{R}[C])\bigstar_{\ell,\{\hat{\mu}_i\}}\bar{\mathcal{B}}(Z_{n,m}(\mathfrak{R}[C]))$, as formal coalgebras, would be incomplete without mentioning their *composition-operations*. These are discussed at length in appendix E on page 117.

THEOREM 1.11. *Let* $\widehat{Z_{n,*}(\mathfrak{R})} = \varprojlim\{p_{m',m}\colon Z_{n,m'}(\mathfrak{R}) \rightarrow Z_{n,m}(\mathfrak{R})\}$. *Then the twisting cochain* $\ell\colon \bar{\mathcal{B}}(Z_{n,m}(\mathfrak{R})) \rightarrow Z_{n,m}(\mathfrak{R})$ *and the* $A(\infty)$*-module structure* $\hat{a}_k\colon Y_{n,m}(\mathfrak{R}) \otimes Z_{n,m}(\mathfrak{R}) \otimes \cdots \otimes Z_{n,m}(\mathfrak{R}) \rightarrow Y_{n,m}(\mathfrak{R})$ *extends to the inverse limit* $\bar{\mathcal{B}}(\widehat{Z_{n,*}(\mathfrak{R})}) = \varprojlim \bar{\mathcal{B}}(Z_{n,m}(\mathfrak{R}))$. $\ell\colon \bar{\mathcal{B}}(\widehat{Z_{n,*}(\mathfrak{R})}) \rightarrow \widehat{Z_{n,*}(\mathfrak{R})}$, $\hat{a}_k\colon \widehat{Y_{n,*}(\mathfrak{R})} \otimes \widehat{Z_{n,m}(\mathfrak{R})} \otimes \cdots \otimes \widehat{Z_{n,m}(\mathfrak{R})} \rightarrow \widehat{Y_{n,*}(\mathfrak{R})}$. *Here* $\widehat{Y_{n,*}(\mathfrak{R})} = \varprojlim q_{m',m}\colon Y_{n,m'}(\mathfrak{R}) \rightarrow Y_{n,m}(\mathfrak{R})$. *In addition there is a canonical isomorphism of twisted tensor products*

$$\widehat{Y_{n,*}(\mathfrak{R})}\bigstar_{\ell,\{\hat{a}_i\}}\bar{\mathcal{B}}(\widehat{Z_{n,*}(\mathfrak{R})}) \cong \widehat{Y_{n,*}(\mathfrak{R})\bigstar_{\ell,\{\hat{\mu}_i\}}\bar{\mathcal{B}}(Z_{n,*}(\mathfrak{R}))}$$

Here, $\widehat{Y_{n,*}(\mathfrak{R})\bigstar_{\ell,\{\hat{\mu}_i\}}\bar{\mathcal{B}}(Z_{n,*}(\mathfrak{R}))}$ *denotes the inverse limit*

$$\varprojlim q_{m',m}\otimes p_{m',m}\colon Y_{n,m'}(\mathfrak{R})\bigstar_{\ell,\{\bar{\mu}_i\}}\bar{\mathcal{B}}(Z_{n,m'}(\mathfrak{R})) \rightarrow Y_{n,m}(\mathfrak{R})\bigstar_{\ell,\{\bar{\mu}_i\}}\bar{\mathcal{B}}(Z_{n,m}(\mathfrak{R}))$$

PROOF. It suffices to verify that the projection-maps in the inverse systems commute with the

1. the twisting cochains:

$$\begin{array}{ccc}
\bar{\mathcal{B}}(Z_{n,m'}(\mathfrak{R})) & \xrightarrow{\ell} & Z_{n,m'}(\mathfrak{R}) \\
\Big\downarrow{\scriptstyle \bar{\mathcal{B}}(p_{m',m})} & & \Big\downarrow{\scriptstyle p_{m',m}} \\
\bar{\mathcal{B}}(Z_{n,m}(\mathfrak{R})) & \xrightarrow{\ell} & Z_{n,m}(\mathfrak{R})
\end{array}$$

2. the structure-maps of the $A(\infty)$-module structure

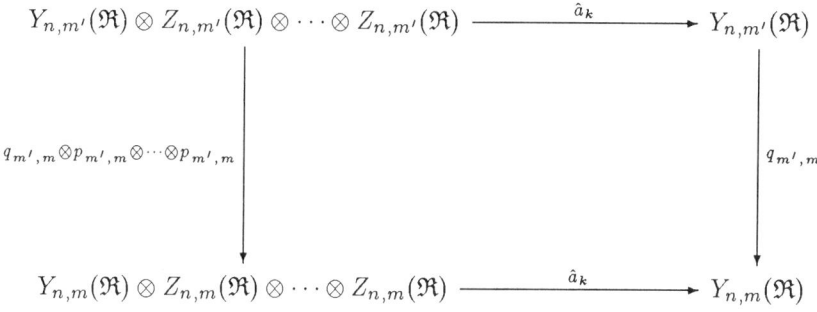

$\square$

PROPOSITION 1.12. *The twisting cochain* $\hat{\ell} \colon \bar{B}(\widehat{Z_{n,*}(\mathfrak{R})}) \to \widehat{Z_{n,*}(\mathfrak{R})}$ *restricts to a twisting cochain* $\hat{\ell} \colon \mathcal{F}(\mathfrak{R}, n) \to \widehat{Z_{n,*}(\mathfrak{R})}$.

Now we will complete the determination of an m-structure for $C \otimes_\ell \mathcal{F}(C)$. Suppose $(C, \{f[C]_n \colon C \to \operatorname{Hom}_{\mathbb{Z}S_n}(\mathfrak{R}[C]_n, C^n)\})$ is a weakly-coherent m-coalgebra. Regard $\mathrm{R}(S_n)$ as a comodule over itself, via the coproduct mapping. We want a map $u_{n,m} \colon \mathrm{R}(S_n) \to Y_{n,m}(\mathfrak{R}[C]) \bigstar_{\ell, \{\hat{\mu}_i\}} \bar{B}(Z_{n,m}(\mathfrak{R}[C]))$ that is a morphism of comodules over $\mathcal{F}(\mathfrak{R}[C], n)$.

We will accomplish this by defining $u_{n,m}$ to be of the form $c_{n,m} \otimes (p_m \circ h_n) \circ \Delta_R$, where

1. $c_{n,m} \colon \mathrm{R}(S_n) \to Y_{n,m}(\mathfrak{R}[C])$ is a suitable map (to be computed later);
2. $h_n \colon \mathrm{R}(S_n) \to \mathcal{F}(\mathfrak{R}[C], n)$ is the morphism of coalgebras constructed in 2.28 on page 56,
3. $p_m \colon \bar{B}(\widehat{Z_{n,*}(\mathfrak{R}[C])}) \to \bar{B}(Z_{n,m}(\mathfrak{R}[C]))$ is the canonical map from the inverse limit to one element in the inverse system.
4. $\Delta_R \colon \mathrm{R}(S_n) \to \mathrm{R}(S_n) \otimes \mathrm{R}(S_n)$ is the coproduct.

PROPOSITION 1.13. *Let* $\mathfrak{R}$ *be an f-resolution equipped with the structure of an* $A(\infty)$-*coalgebra. The maps* $u_{n,m} = c_{n,m} \otimes (p_m \circ h_n) \circ \Delta_R \colon \mathrm{R}(S_n) \to Y_{n,m}(\mathfrak{R}) \bigstar_{\ell, \{\hat{\mu}_i\}} \bar{B}(Z_{n,m}(\mathfrak{R}))$, *as defined above, are morphisms of comodules. Here* $\Delta_R \colon \mathrm{R}(S_n) \to \mathrm{R}(S_n) \otimes \mathrm{R}(S_n)$ *is the standard coproduct.*

PROOF. We must show that $(1 \otimes \Delta_{\mathfrak{R}[C]}) \circ u_{n,m} = u_{n,m} \otimes (p_m \circ h_n) \circ \Delta_R$. If we plug the definition of u_n into this we get $((c_{n,m} \otimes (p_m \circ h_n)) \circ \Delta_R) \otimes h_n \circ \Delta_R = c_{n,m} \otimes (p_m \circ h_n) \otimes (p_m \circ h_n) \circ \Delta_R \otimes 1 \circ \Delta_R = c_{n,m} \otimes (p_m \circ h_n) \otimes h_n \circ 1 \otimes \Delta_R \circ \Delta_R$ (by the co-associativity of Δ_R) $= c_{n,m} \otimes ((p_m \circ (p_m \circ h_n)) \otimes (p_m \circ h_n) \circ \Delta_R) \circ \Delta_R = c_{n,m} \otimes (\Delta_{\mathfrak{R}[C]} \circ (p_m \circ h_n)) \circ \Delta_R$ (because h_n is a morphism of coalgebras) $= (1 \otimes \Delta_{\mathfrak{R}[C]}) \circ u_n$. $\square$

For any sequence, α, of length n of nonnegative integers let $y_\alpha \colon \mathrm{R}(S_n) \to \mathfrak{R}_{|\alpha|+n}$ be maps of degree $|\alpha|$ and define $c_n(\alpha) = \downarrow^{|\alpha|} \circ y_\alpha \colon \mathrm{R}(S_n) \to Y(\mathfrak{R}, \alpha)$ and $c_{n,m} \colon \mathrm{R}(S_n) \to Y_{n,m}(\mathfrak{R})$ to be

$$\sum_{\substack{\alpha \\ |\alpha| \leq m}} c_n(\alpha)$$

The requirement that the m-structure of $C \otimes_\ell \mathcal{F}(C)$ project to that of C implies that $y_0 =$ the structure map of C, where 0 in this context represents the sequence of n zeroes.

These requirements and the formula for the boundary in 1.9 on page 77, gives the following formula for the $\{y_\alpha\}$:

PROPOSITION 1.14. *Let* $(C, \{\mathfrak{f}[C]_n : C \rightarrow \mathrm{Hom}_{\mathbb{Z}S_n}(\mathfrak{R}[C]_n, C^n)\})$ *be a weakly-coherent m-coalgebra with associated canonical* $A(\infty)$-*coalgebra structure* $\{\Delta_i : C \rightarrow C^i\}$ *(see see 1.5 on page 36). In addition, suppose* $\mathfrak{R}[C]$ *is the coordinate coalgebra of* $\mathcal{F}(C)$ *with coalgebra structure* $\Delta_{\mathfrak{R}[C]} : \mathfrak{R}[C] \rightarrow \mathfrak{R}[C] \otimes \mathfrak{R}[C]$. *The* $\{y_\alpha\}$, *defined above must satisfy the following formula:*

$$y_\alpha \circ \partial_R - (-1)^{|\alpha|} \partial_{\mathfrak{R}} \circ y_\alpha$$

$$- \sum_{t=1}^{|\alpha|} \sum_{i=2}^{\infty} (-1)^{|\alpha|i+(i+1)(t+n)} \Delta_i \circ_{t+n} y_\alpha(t+n, i) +$$

$$\sum_{i=2}^{\infty} \sum_{t=2}^{n} (-1)^{|\alpha|i} \mathcal{Z}\{(t, i-1), (n-t+\sum_{j=1}^{t-1} \alpha_j, \sum_{j=t}^{n} \alpha_j)\} \Delta_i \circ_t y_\alpha(t, i)$$

$$- \sum_{i=2}^{\infty} (-1)^{i|\alpha|} \sum_{\beta_1 + \cdots + \beta_i = \alpha} (-1)^{\sum_{j=2}^{i} |\beta_j|(|\beta_1| + \sum_{j=2}^{i-1} |\beta_j| - 1)} \mathcal{S}_n(\mathcal{Z}\{\beta_1, \ldots, \beta_i\}) = 0$$

$$\circ_{t_i} \circ y_{\beta_1} \otimes z_{\beta_2} \otimes \cdots \otimes z_{\beta_i} \circ \Delta_{\mathfrak{R}[C]}^{i-1}$$

Remarks.   1.14.1. This formula allows us to calculate the $\{y_\alpha\}$, given contracting chain-homotopies for the $Y_{n,m}(\mathfrak{R})$ — it is particularly suited to machine-computation. It expresses the boundary of a given y_α in terms of the $\{z_\alpha\}$ and other $y_{\alpha'}$, where α' is lexicographically less than α. The fact that $Y_{n,m}(\mathfrak{R})$ is acyclic in positive dimensions implies that we can do an inductive computation of the $\{y_\alpha\}$.

1.14.2. The $y_\alpha(u, v)$ are defined as follows (compare remark 2.27.2 on page 55): Let $\gamma(i)$ be defined to be the *largest* value of t such that $\sum_{k=1}^{t-1} \alpha_k < i \leq \sum_{k=1}^{t} \alpha_k$. Then $y_\alpha(i, j)$ is defined to be $\begin{cases} 0 & \text{if } \alpha_{\gamma(i+j-1)} \leq j-1 \\ y_{\alpha'} & \text{if } \alpha_{\gamma(i+j-1)} > j-1 \end{cases}$, where α' is the sequence that results from subtracting $j-1$ from the $\gamma(i+j-1)^{\text{th}}$ term of α.

1.14.3. The $\{t_i\}$ are defined in remark 2.14.2 on page 45.

Now we can define the coordinate coalgebra of $C \otimes_\ell \mathcal{F}(C)$:

DEFINITION 1.15. *Let* $\mathfrak{R}$ *be an f-resolution equipped with the structure of an* $A(\infty)$-*coalgebra. The formal coalgebras* $\hat{Y}_n(\mathfrak{R}) \bigstar_{\ell, \{\hat{\mu}_i\}} \mathcal{F}(\mathfrak{R}, n)$ *are acyclic in positive dimensions. The comodule morphism* $u_n : R(S_n) \rightarrow \hat{Y}_n(\mathfrak{R}) \bigstar_{\ell, \{\hat{\mu}_i\}} \mathcal{F}(\mathfrak{R}, n)$, *constructed above, induces a homomorphism* $(u_n)_* : \mathbb{Z} \rightarrow H_0(\hat{Y}_n(\mathfrak{R}) \bigstar_{\ell, \{\hat{\mu}_i\}} \mathcal{F}(\mathfrak{R}, n))$ *and the*

pullback of the diagram

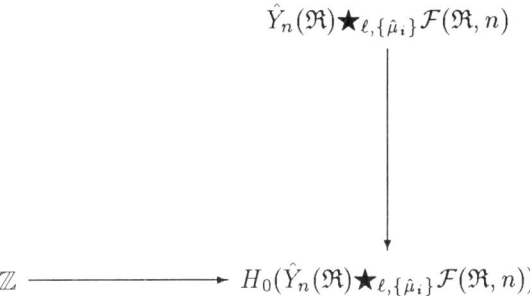

$$\hat{Y}_n(\mathfrak{R})\bigstar_{\ell,\{\hat{\mu}_i\}}\mathcal{F}(\mathfrak{R},n)$$

$$\mathbb{Z} \longrightarrow H_0(\hat{Y}_n(\mathfrak{R})\bigstar_{\ell,\{\hat{\mu}_i\}}\mathcal{F}(\mathfrak{R},n))$$

is an f-resolution denoted $\mathcal{P}(\mathfrak{R}, n)$. This is defined to be the coordinate coalgebra of the m-structure of $C \otimes_\ell \mathcal{F}(C)$.

Remarks. 1.15.1. $\mathcal{P}(\mathfrak{R}, n)$ comes equipped with a natural projection $\mathcal{P}(\mathfrak{R}, n) \to \mathfrak{R}$, and is a right comodule over $\mathcal{F}(\mathfrak{R}, n)$.

1.15.2. Note that $\mathcal{P}(\mathfrak{R}, n)$ is a strict left m-module over $\mathcal{F}(\mathfrak{R}, n)$ — see 3.14 on page 27.

1.15.3. It is not hard to see that $c_{n,m} \colon R(S_n) \to \hat{Y}_n(\mathfrak{R})$ is a chain-map in dimension 0. This implies that we can carry out a construction similar to the above and get $\mathcal{Y}_n(\mathfrak{R}) \subset \hat{Y}_n(\mathfrak{R})$ with the property that $H_0(\mathcal{Y}_n(\mathfrak{R})) = \mathbb{Z}$, and $\mathcal{P}(\mathfrak{R}, n) = \mathcal{Y}_n(\mathfrak{R})\bigstar_{\ell,\{\hat{\mu}_i\}}\mathcal{F}(\mathfrak{R}, n)$.

We can summarize the results of this section so far in:

THEOREM 1.16. *Let* $(C, \{\mathfrak{f}[C]_n \colon C \to \operatorname{Hom}_{\mathbb{Z}S_n}(\mathfrak{R}[C]_n, C^n)\})$ *be a weakly coherent m-coalgebra. Let* $e_n \colon \mathcal{P}(\mathfrak{R}[C], n) \to \hat{Y}_n(\mathfrak{R}[C])\bigstar_{\ell,\{\hat{\mu}_i\}}\mathcal{F}(\mathfrak{R}[C], n)$ *be the inclusion. Then* $\mathcal{A}(\mathfrak{f}[C]_n) = \operatorname{Hom}_{\mathbb{Z}S_n}(e_n, 1) \circ I \circ A(\mathfrak{f}[C]_n)' \colon C \otimes_\ell \mathcal{F}(C) \to \operatorname{Hom}_{\mathbb{Z}S_n}(\mathcal{P}(\mathfrak{R}[C], n), (C \otimes_\ell \mathcal{F}(C))^n)$ *is the structure map of a coherent m-structure on* $C \otimes_\ell \mathcal{F}(C)$. *Here* $I \colon \operatorname{Hom}_{\mathbb{Z}S_n}(\hat{Y}_n(\mathfrak{R}[C])\bigstar_{\ell,\{\hat{\mu}_i\}}\mathcal{F}(\mathfrak{R}[C], n), *) \cong \operatorname{Hom}_{\mathbb{Z}S_n}(Y_{n,m}(\mathfrak{R}[C])\bigstar_{\ell,\{\hat{\mu}_i\}}\bar{B}(Z_{n,m}(\mathfrak{R}[C])), *)$ *is the canonical isomorphism (which exists because* $\hat{Y}_n(\mathfrak{R}[C])\bigstar_{\ell,\{\hat{a}_i\}}\bar{B}(\widehat{Z_{n,*}(\mathfrak{R}[C])}) \cong \varprojlim Y_{n,m}(\mathfrak{R}[C])\bigstar_{\ell,\{\hat{a}_i\}}\bar{B}(Z_{n,m}(\mathfrak{R}[C]))$, *the inverse limit of the inverse system*

$$q_{m',m} \otimes p_{m',m} \colon Y_{n,m'}(\mathfrak{R}[C])\bigstar_{\ell,\{\hat{a}_i\}}\bar{B}(Z_{n,m'}(\mathfrak{R}[C]))$$
$$\to Y_{n,m}(\mathfrak{R}[C])\bigstar_{\ell,\{\hat{a}_i\}}\bar{B}(Z_{n,m}(\mathfrak{R}[C]))$$

as proved in 1.11 on page 80).

REMARK. 1.16.1. It is clear that the map in question is a chain-map: see 1.7 on page 74 for a definition of $\mathcal{A}(\mathfrak{f}[C]_n)'$ and a proof that it is a chain-map.

At first glance it may appear that we haven't accomplished much with all of this work — we have only found a weakly coherent m-structure on $C \otimes_\ell \mathcal{F}(C)$, which is an acyclic complex. This turns out, however, to be the basis for all twisted tensor products with C as the base.

In order to discuss general twisted tensor products, we must first recall the concept of cotensor products over a coalgebra:

DEFINITION 1.17. Let:

1. A be a DGA-coalgebra;
2. M be a right A-comodule and;
3. N be a left A-comodule

Recall that this means these are DGA modules equipped with maps:

1. $\Delta_M \colon M \to M \otimes A$;
2. $\Delta_N \colon N \to A \otimes N$;satisfying the identities: $(\Delta_M \otimes 1) \circ \Delta_M = (1 \otimes \Delta_A) \circ \Delta_M$ and $(1 \otimes \Delta_N) \circ \Delta_N = (\Delta_A \otimes 1) \circ \Delta_N$, where $\Delta_A \colon A \to A \otimes A$ is the coproduct.

The *cotensor product* of M and N over A, denoted $M \otimes^A N$, is defined to be the kernel of the map $\Delta_M \otimes 1_N - 1_M \otimes \Delta_N \colon M \otimes N \to M \otimes A \otimes N$.

REMARK. 1.17.1. The cotensor product has the following well-known properties (see [10]):

If A is a DGA coalgebra, $\xi \colon A \to F$ is a twisting cochain, and Z is a right A-comodule then $Z \otimes^A (A \otimes_\xi F) = Z \otimes_\xi F$. In the twisted tensor product $Z \otimes_\xi F$ the differential is defined with respect to the comodule structure of Z: $\partial_\xi = \partial_\otimes + (1 \otimes \mu) \circ (1 \otimes \xi \otimes 1) \circ s \otimes 1$, where $s \colon Z \to Z \otimes A$ defines the comodule structure of Z. It is also easy to see that the corresponding statement holds for our $\bigstar$-twisted tensor product:

$$(F \bigstar_\xi A) \otimes^A B = F \bigstar_\xi B$$

The following is clear:

PROPOSITION 1.18. *Under the hypotheses of 1.17 above, suppose that M, N, and A are formal coalgebras, and that the structure maps Δ_M, Δ_N, are morphisms of formal coalgebras. Then the cotensor product, $M \otimes^A N$, is also a formal coalgebra.*

The remainder of this section will be spent in computing m-coalgebra structures of general twisted tensor products, using the computation of the m-coalgebra structure of $C \otimes_\ell \mathcal{F}(C)$. We begin with the simplest and most clear-cut case:

PROPOSITION 1.19. *Let C and F weakly coherent m-coalgebras with structure maps $(C, \{\mathfrak{f}[C]_n \colon C \to \operatorname{Hom}_{\mathbb{Z}S_n}(\mathfrak{R}[C]_n, C^n)\})$ and $(F, \{\mathfrak{f}[F]_n \colon F \to \operatorname{Hom}_{\mathbb{Z}S_n}(\mathfrak{R}[F]_n, F^n)\})$, respectively. Suppose that F is a strict left m-module over $\mathcal{F}(C)$ with action-map $a \colon \mathcal{F}(C) \otimes F \to F$, and corresponding maps of coordinate coalgebras*

$$\{r_n \colon \mathfrak{R}[F]_n \to \mathcal{F}(\mathfrak{R}[C], n) \otimes \mathfrak{R}[F]_n\}$$

Then there exists a weakly-coherent m-structure on the twisted tensor product $C \otimes_\xi F = (C \otimes_\ell \mathcal{F}(C)) \otimes_{\mathcal{F}(C)} F$ with coordinate coalgebras $\{\mathcal{P}(\mathfrak{R}[C], n) \otimes^{\mathcal{F}(\mathfrak{R}[C], n)} \mathfrak{R}[F]_n = \mathcal{Y}_n(\mathfrak{R}[C]) \bigstar_{\{\ell, \hat{m}\}} \mathfrak{R}[F]_n\}$.

The structure map of $C \otimes_\xi F$ is the composite $\operatorname{Hom}_{\mathbb{Z}S_n}(\iota_n, a_n) \circ j_n \circ (\mathcal{A}(\mathfrak{f}[C]_n)' | C \otimes 1) \otimes \mathfrak{f}[F]_n$, where

1. *$\mathcal{A}(\mathfrak{f}[C]_n) \colon C \otimes_\ell \mathcal{F}(C) \to \operatorname{Hom}_{\mathbb{Z}S_n}(\mathcal{P}(\mathfrak{R}[C], n), (C \otimes_\ell \mathcal{F}(C))^n)$ was defined in 1.16 on page 83;*

2.

$$j_n \colon \mathrm{Hom}_{\mathbb{Z}S_n}(\mathcal{P}(\mathfrak{R}[C], n), (C \otimes_\ell \mathcal{F}(C))^n) \otimes \mathrm{Hom}_{\mathbb{Z}S_n}(\mathfrak{R}[F]_n, F^n)$$
$$\to \mathrm{Hom}_{\mathbb{Z}S_n}(\mathcal{P}(\mathfrak{R}[C], n) \otimes \mathfrak{R}[F]_n, (C \otimes_\ell \mathcal{F}(C))^n \otimes F^n)$$

is the natural map;

3. $a_n \colon \mathcal{F}(C)^n \otimes F^n \to F^n$ *is induced by the action of $\mathcal{F}(C)$ on F;*
4. $\iota_n \colon \mathcal{Y}(\mathfrak{R}[C]) \bigstar_{\{\ell, \hat{\mu}_i\}} \mathfrak{R}[F]_n \to \mathcal{P}(\mathfrak{R}[C], n) \otimes \mathfrak{R}[F]_n$ *is the inclusion of the cotensor product.*

REMARK. 1.19.1. The m-structure defined on $C \otimes_\xi F$ by the proposition above will be called the canonical m-structure of $C \otimes_\xi F$. It is clearly natural with respect to strict morphisms of m-coalgebras and strict m-modules. A little reflection shows that the corresponding statement is valid for morphisms that are not strict:

Suppose we are given an elementary morphism $f \colon C' \to C$ that is *not* strict (i.e. it has a strict inverse $h \colon C \to C'$ with $(f, h, \varphi) \colon C' \to C$ a *contraction*, according to the definition in 3.9 on page 25). We must refer to 3.13 on page 26 to conclude that the morphism f induces a morphism of m-Hopf algebras $\mathcal{F}(f) \colon \mathcal{F}(C') \to \mathcal{F}(C)$. We can now define the pullback of a twisted tensor product over f via $\mathcal{F}(f)$: Given a twisting cochain $\xi \colon C \to F$ regard it as the composite $f(\xi) \circ \ell$, where $f(\xi)$ is the classifying map $\mathcal{F}(C) \to F$. Now define $f^*\xi$ to be $f(\xi) \circ \mathcal{F}(f) \circ \ell$. This will still be a twisting cochain since $\mathcal{F}(f)$ is a DGA-algebra homomorphism. The formula in 2.32 on page 58 shows that $f(\xi) \circ \mathcal{F}(f) \circ \ell$ will be *twisted* in some sense, i.e. the diagram

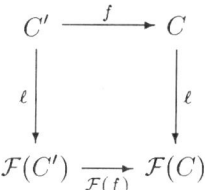

will only commute up to a chain-homotopy. Nevertheless there will exist an elementary morphism $C' \otimes_{f \cdot \xi} F \to C \otimes_\xi F$.

PROOF. Most of this has already been proved. The structure map given above is essentially that of the tensor product $(C \otimes_\ell \mathcal{F}(C)) \otimes F$ composed with the projection $(C \otimes_\ell \mathcal{F}(C)) \otimes F \to (C \otimes_\ell \mathcal{F}(C)) \otimes_{\mathcal{F}(C)} F$. The cotensor product $\{\mathcal{P}(\mathfrak{R}[C], n) \otimes^{\mathcal{F}(\mathfrak{R}[C], n)} \mathfrak{R}[F]_n\}$ is precisely the submodule of $\mathcal{P}(\mathfrak{R}[C], n) \otimes \mathfrak{R}[F]_n$ for which this procedure is well-defined. □

The following result will be very important in constructing algebraic models of homotopy types:

THEOREM 1.20. *Let I denote the unit interval, equipped with its canonical coherent m-structure. Let C and F be weakly-coherent m-coalgebras and let F be a left m-module over $\mathcal{F}(C \otimes I)$ with structure map $A \colon \mathcal{F}(C \otimes I) \otimes F \to F$. Let $k_i \colon \mathcal{F}(C \otimes p_i) \to \mathcal{F}(C \otimes I)$, $i = 0, 1$, be induced by the inclusions of the endpoints, and let $a_i = A \circ k_i \otimes 1 \colon \mathcal{F}(C) \otimes F \to F$. Then there exists an morphism of weakly-coherent m-coalgebras: $w \colon C \otimes_{a_1} F \to C \otimes_{a_1} F$, whose underlying map is an isomorphism of chain-complexes. This map is given by: $w = 1 + (A \circ \mathfrak{h}) \cap *$, where $\mathfrak{h} \colon C \to C \otimes I$ sends $c \to c \otimes q$.*

Remarks.   1.20.1. Regard $A \circ \mathfrak{h}\colon C \to F$ as a cochain on C in the formula above (this means we use the coproduct of C in evaluating the cap-product).

1.20.2. We regard the unit interval as being defined by $I_0 = \mathbb{Z} \oplus \mathbb{Z}$, generated by the endpoints p_0 and p_1, and $I_1 = \mathbb{Z}$ generated by q, where $\partial q = p_1 - p_0$. The canonical m-coalgebra structure on I is defined by regarding it as a 1-simplex.

PROOF. The map $\mathfrak{h}\colon C \to C \otimes I$ can be regarded as a chain-homotopy between the inclusions $C \otimes p_i \to C \otimes I$. The map, w, can be regarded as $\partial_A \circ \mathfrak{h} \otimes 1 + \mathfrak{h} \otimes 1 \circ \partial_{C \otimes_{a_0} F}$, where ∂_A is the (twisted) differential of $C \otimes I \otimes_A F$. It follows that $C \otimes I \otimes_A F$ is the algebraic mapping cylinder of w. $\cdot \square$

2. Approximate m-structures on twisted tensor products

This section will consider the question: Suppose F is a non-strict m-module over $\mathcal{F}(C)$ (this situation actually occurs in applications). It turns out that there is a well-defined m-structure like that in 1.19 on page 84 except that:

1. It might not turn out to be weakly-coherent;
2. The projection $(C \otimes_\ell \mathcal{F}(C)) \otimes F \to (C \otimes_\ell \mathcal{F}(C)) \otimes_{\mathcal{F}(C)} F$ might not be a morphism of m-structures — it might only preserve m-structures up to a homotopy.

Note that, since m-structures are only preserved up to a chain-homotopy in m-modules, it is necessary to verify that certain homological-algebraic constructs are well-defined. In order to do this we need the following simple lemma:

LEMMA 2.1. *Consider the following diagram of chain-complexes and maps*

$$
\begin{array}{ccc}
C & \xrightarrow{\ f_2\ } & D \\
{\scriptstyle d_1}\uparrow & & \uparrow{\scriptstyle x} \\
A & \xrightarrow{\ f_1\ } & B
\end{array}
$$

where all maps represented by solid arrows are given and the horizontal arrows represent split surjections. A map $x\colon B \to D$ making this diagram commutative up to a chain-homotopy exists if and only if $f_2 \circ d_1 | \ker f_1$ is nullhomotopic. If this condition is satisfied, then x is uniquely determined up to a chain-homotopy.

PROOF. Clearly, if the diagram is homotopy-commutative, then the restriction of the chain-homotopy to $\ker f_1$ will equal $f_2 \circ d_1 | \ker f_1$, since $x \circ f_1 | \ker f_1 = 0$. Conversely, if $f_2 \circ d_1 | \ker f_1$ is nullhomotopic, let $\varphi\colon \ker f_1 \to D$ be a nullhomotopy. The fact that f_1 is a split surjection implies that $\ker f_1$ is a direct summand of A, and φ extends to a homotopy $\Phi\colon A \to D$. Clearly, $G = f_2 \circ d_1 - \partial \Phi$ vanishes on $\ker f_1$ so G factors through f_1 and gives rise to x. We follow the convention that $\partial \Phi = \partial \circ \Phi - (-1)^{deg(\Phi)}\Phi \circ \partial$. The resulting diagram is clearly homotopy-commutative. We may vary φ on $\ker f_1$ or the extension to A, but the resulting homotopy Φ' must have the property that $\partial \Phi' | \ker f_1 = \partial \Phi | \ker f_1$. The resulting map, x, will have varied by a chain-homotopy, namely $\partial(\Phi' - \Phi)|B$, where we identify A with $B \oplus \ker f_1$. $\square$

DEFINITION 2.2. Consider the following diagram of maps of DGA-modules:

1. $C_1 \otimes A \xleftarrow{\ f_1\ } P_1 \xrightarrow{\ g_1\ } C_1$

2. $A \otimes C_2 \xleftarrow{\ f_2\ } P_2 \xrightarrow{\ g_2\ } C_2$

The *generalized cotensor product* of these two diagrams, denoted $[P_1 \otimes^A P_2]$, is defined to be the kernel of the map $f_1 \otimes g_2 - g_1 \otimes f_2 \colon P_1 \otimes P_2 \to C_1 \otimes A \otimes C_2$. There exist natural maps:

1. $g_1 \otimes g_2 \colon [P_1 \otimes^A P_2] \to C_1 \otimes C_2$;
2. $1 \otimes g_2 \colon [P_1 \otimes^A P_2] \to P_1 \otimes C_2$;
3. $g_1 \otimes 1 \colon [P_1 \otimes^A P_2] \to C_1 \otimes P_2$.

REMARK. 2.2.1. Note that if the maps $\{g_i\}$ are both the identity map this reduces to the usual cotensor product.

PROPOSITION 2.3. *Let* $(M, \{\mathfrak{f}[M]_n \colon M \to \mathrm{Hom}_{\mathbb{Z}S_n}(\mathfrak{R}[M]_n, M^n)\})$ *be a right m-module over a strict m-Hopf algebra* $(A, \mu[A], \{\mathfrak{f}[A]_n \colon A \to \mathrm{Hom}_{\mathbb{Z}S_n}(\mathfrak{R}[A]_n, A^n)\})$, *and let* $(F, \{\mathfrak{f}[F]_n \colon F \to \mathrm{Hom}_{\mathbb{Z}S_n}(\mathfrak{R}[F]_n, F^n)\})$ *be a left m-module over* A. *Suppose that the underlying chain-complexes of* M, A, F, *and* $M \otimes_A F$ *are all free. Then the tensor product* $M \otimes_A F$ *can be equipped with an m-structure* $f_x \colon M \otimes_A F \to \mathrm{Hom}_{\mathbb{Z}S_n}([\mathfrak{R}[P_M] \otimes^{\mathfrak{R}[A]} \mathfrak{R}[P_F]], (M \otimes_A F)^n)$, *such that the diagram in figure 4.4.1 (on page 88) commutes up to a chain-homotopy. Here* $p \colon M \otimes F \to M \otimes_A F$ *is the projection that identifies the right action of* A *on* M *with the left action of* A *on* F, *and* $j \colon [\mathfrak{R}[P_M] \otimes^{\mathfrak{R}[A]} \mathfrak{R}[P_F]] \to \mathfrak{R}[M] \otimes \mathfrak{R}[F]$ *is the natural map of the generalized cotensor product. This m-structure is uniquely determined up to a chain-homotopy. Let* $(\iota_M, \varphi_M, r_M, g_M) \colon M \otimes A \to PM \to M$ *and* $(\iota_F, \varphi_F, r_F, g_F) \colon A \otimes F \to PF \to F$ *denote the structure morphisms of* M *and* F, *respectively, and let the m-structures of* P_F *and* P_M *be given by* $(P_F, \{\mathfrak{f}[P_F]_n \colon P_F \to \mathrm{Hom}_{\mathbb{Z}S_n}(\mathfrak{R}[P_F]_n, P_F^n)\})$ *and* $(P_M, \{\mathfrak{f}[P_M]_n \colon P_M \to \mathrm{Hom}_{\mathbb{Z}S_n}(\mathfrak{R}[P_M]_n, P_M^n)\})$. *Then this m-structure is given by:*

$$\mathrm{Hom}_{\mathbb{Z}S_n}(j, p^n) \circ \mathfrak{f}[M \otimes F] \circ t - \partial\{\mathrm{Hom}_{\mathbb{Z}S_n}(w_1, p^n) \circ V_1 \circ \Phi_1 \otimes \mathfrak{f}[F] \circ z\} \circ t$$
$$+ \partial\{\mathrm{Hom}_{\mathbb{Z}S_n}(w_2, p^n) \circ V_2 \circ \mathfrak{f}[F] \otimes \Phi_2 \circ z\} \circ t$$
$$\colon M \otimes_A F \to \mathrm{Hom}_{\mathbb{Z}S_n}([\mathfrak{R}[P_M] \otimes^{\mathfrak{R}[A]} \mathfrak{R}[P_F]], (M \otimes_A F)^n)$$

where:

1. $t \colon M \otimes_A F \to M \otimes F$ *is a right-inverse to* p *(not necessarily a chain-map)*;
2. $z \colon M \otimes F \to M \otimes_A \otimes F$ *is a right-inverse to* $a_M \otimes 1 - 1 \otimes a_F \colon M \otimes A \otimes F \to M \otimes F$ *(not necessarily a chain-map)*;
3. $\Phi_1 = r_M^n \circ \mathfrak{f}[P_M] \circ \varphi_M \circ \iota_M \colon M \otimes A \to \mathrm{Hom}_{\mathbb{Z}S_n}(\mathfrak{f}[P_M], M^n)$, *the chain-homotopy between* $\mathfrak{f}[M] \circ a_M$ *and* $\mathrm{Hom}_{\mathbb{Z}S_n}(1, a_M^n) \circ \mathfrak{f}[M \otimes A]$, *where* $a_M \colon M \otimes A \to M$ *represents the action of* A *on* M;
4. $\Phi_2 = r_F^n \circ \mathfrak{f}[P_F] \circ \varphi_F \circ \iota_F \colon A \otimes F \to \mathrm{Hom}_{\mathbb{Z}S_n}(\mathfrak{R}[P_F]_n, F^n)$, *the chain-homotopy between* $\mathfrak{f}[F] \circ a_F$ *and* $\mathrm{Hom}_{\mathbb{Z}S_n}(1, a_F^n) \circ \mathfrak{f}[A \otimes F]$, *where* $a_F \colon A \otimes F \to F$ *represents the action of* A *on* F.
5. *We require that*

$$\mathrm{Hom}_{\mathbb{Z}S_n}(w_1, p^n) \circ V_1 \circ \Phi_1 \otimes \mathfrak{f}[F] - \mathrm{Hom}_{\mathbb{Z}S_n}(w_2, p^n) \circ V_2 \circ \mathfrak{f}[M] \otimes \Phi_2$$
$$| \ker(a_M \otimes 1 - 1 \otimes a_F) = 0$$
$$\colon M \otimes A \otimes F \to \mathrm{Hom}_{\mathbb{Z}S_n}([\mathfrak{R}[P_M] \otimes^{\mathfrak{R}[A]} \mathfrak{R}[P_F]]_n, (M \otimes F)^n)$$

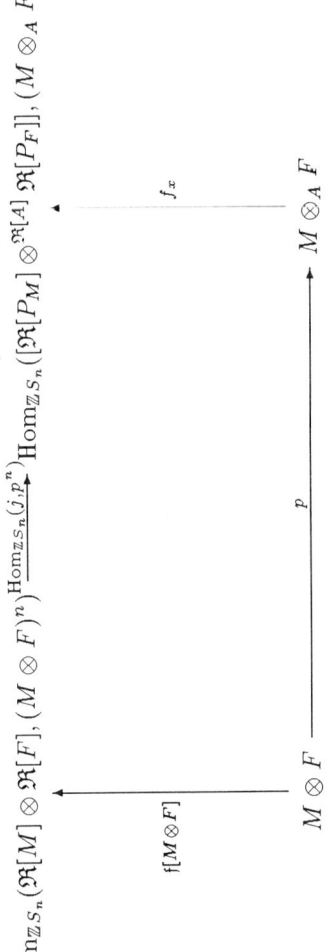

Figure 4.4.1.

Here

$$V\colon \operatorname{Hom}_{\mathbb{Z}S_n}(\mathfrak{R}[M]_n, M^n) \otimes \operatorname{Hom}_{\mathbb{Z}S_n}(\mathfrak{R}[F]_n, F^n)$$
$$\to \operatorname{Hom}_{\mathbb{Z}S_n}(\mathfrak{R}[M]_n \otimes \mathfrak{R}[F]_n, (M \otimes F)^n)$$

$$V_1\colon \operatorname{Hom}_{\mathbb{Z}S_n}(\mathfrak{R}[P_M]_n, M^n) \otimes \operatorname{Hom}_{\mathbb{Z}S_n}(\mathfrak{R}[F]_n, F^n)$$
$$\to \operatorname{Hom}_{\mathbb{Z}S_n}(\mathfrak{R}[P_M] \otimes \mathfrak{R}[F], (M \otimes F)^n)$$

$$V_2\colon \operatorname{Hom}_{\mathbb{Z}S_n}(\mathfrak{R}[M], M^n) \otimes \operatorname{Hom}_{\mathbb{Z}S_n}(\mathfrak{R}[P_F]_n, F^n)$$
$$\to \operatorname{Hom}_{\mathbb{Z}S_n}(\mathfrak{R}[M] \otimes \mathfrak{R}[P_F], (M \otimes F)^n)$$

$$w_1\colon [\mathfrak{R}[P_M] \otimes^{\mathfrak{R}[A]} \mathfrak{R}[P_F]] \to \mathfrak{R}[P_M] \otimes \mathfrak{R}[F]$$

$$w_2\colon [\mathfrak{R}[P_M] \otimes^{\mathfrak{R}[A]} \mathfrak{R}[P_F]] \to \mathfrak{R}[M] \otimes \mathfrak{R}[P_F]$$

are the natural maps.

Remarks.   2.3.1. Note that, if M and F are strict A-modules (see 3.14 on page 27), then the m-structure on the tensor product reduces to $\operatorname{Hom}_{\mathbb{Z}S_n}(j, p^n) \circ \mathfrak{f}[M \otimes F] \circ t$.

2.3.2. The map $t\colon M \otimes_A F \to M \otimes F$ is not, in general, a chain-map — it is only a lift of elements of $M \otimes_A F \to M \otimes F$. The fact that it isn't a chain map doesn't matter since the map it is composed with annihilates the kernel of p.

2.3.3. We follow the convention that $\partial\Phi = \partial \circ \Phi - (-1)^{\deg(\Phi)}\Phi \circ \partial$.

2.3.4. The term " free" is deliberately a bit vague. In the present context is simply means "$\mathbb{Z}$-free". The results of this paper can be generalized to non-simply connected spaces, in which case "free" means "free $\mathbb{Z}\pi$-chain complexes, where π is the fundamental group". The condition of freeness is used to guarantee the existence of the map L.

PROOF. This is an application of lemma 2.1 on page 86. Note that the kernel of p is the image of $M \otimes A \otimes F$ under the mapping $a_M \otimes 1 - 1 \otimes a_F\colon M \otimes A \otimes F \to M \otimes F$
Claim:

$$\operatorname{Hom}_{\mathbb{Z}S_n}(j, p^n) \circ V \circ \operatorname{Hom}_{\mathbb{Z}S_n}(1, a_M^n) \otimes 1 \circ \mathfrak{f}[M \otimes A] \otimes \mathfrak{f}[F]$$
$$- \operatorname{Hom}_{\mathbb{Z}S_n}(j, p^n) \circ V \circ 1 \otimes \operatorname{Hom}_{\mathbb{Z}S_n}(1, a_F^n) \circ \mathfrak{f}[M] \otimes \mathfrak{f}[F \otimes A] = 0$$
$$\colon M \otimes A \otimes F \to \operatorname{Hom}_{\mathbb{Z}S_n}([\mathfrak{R}[P_M] \otimes^{\mathfrak{R}[A]} \mathfrak{R}[P_F]]_n, (M \otimes_A F)^n)$$

The quantity in the parentheses is just the result of, first taking the *product* m-coalgebra structure $\mathfrak{f}[M \otimes A \otimes F]$ (see 3.6 on page 23) and then, multiplying, respectively, the M-factor and the F-factor by the A-factor and subtracting the results. But these products are identified by p.

Now the definition of Φ_1 above, and 3.12 on page 25 implies that

$$\operatorname{Hom}_{\mathbb{Z}S_n}(j, p^n) \circ V \circ \partial\Phi_1 \otimes \mathfrak{f}[F] = \operatorname{Hom}_{\mathbb{Z}S_n}(j, p^n) \circ \mathfrak{f}[M \otimes F] \circ a_M \otimes 1$$
$$- \operatorname{Hom}_{\mathbb{Z}S_n}(j, p^n) \circ V \circ \operatorname{Hom}_{\mathbb{Z}S_n}(1, a_M^n) \otimes 1 \circ \mathfrak{f}[M \otimes A] \otimes \mathfrak{f}[F]$$
$$\colon M \otimes A \otimes F \to \operatorname{Hom}_{\mathbb{Z}S_n}([\mathfrak{R}[P_M] \otimes^{\mathfrak{R}[A]} \mathfrak{R}[P_F]]_n, (M \otimes_A F)^n)$$

Similarly, the definition of Φ_2 implies that

$$\operatorname{Hom}_{\mathbb{Z}S_n}(j, p^n) \circ V \circ \mathfrak{f}[M] \otimes \partial\Phi_2 = \operatorname{Hom}_{\mathbb{Z}S_n}(j, p^n) \circ \mathfrak{f}[M \otimes F] \circ 1 \otimes a_F$$
$$- \operatorname{Hom}_{\mathbb{Z}S_n}(j, p^n) \circ V \circ 1 \otimes \operatorname{Hom}_{\mathbb{Z}S_n}(1, a_F^n) \circ \mathfrak{f}[M] \otimes \mathfrak{f}[A \otimes F]$$
$$: M \otimes A \otimes F \rightarrow \operatorname{Hom}_{\mathbb{Z}S_n}([\mathfrak{R}[P_M] \otimes^{\mathfrak{R}[A]} \mathfrak{R}[P_F]]_n, (M \otimes_A F)^n)$$

This and the claim above imply that

$$\operatorname{Hom}_{\mathbb{Z}S_n}(j, p^n) \circ \mathfrak{f}[M \otimes F] \circ (a_M \otimes 1 - 1 \otimes a_F) =$$
$$\operatorname{Hom}_{\mathbb{Z}S_n}(j, p^n) \circ V \circ (\partial\Phi_1 \otimes \mathfrak{f}[F] - \mathfrak{f}[M] \otimes \partial F2): M \otimes A \otimes F \rightarrow$$
$$\operatorname{Hom}_{\mathbb{Z}S_n}([\mathfrak{R}[P_M] \otimes^{\mathfrak{R}[A]} \mathfrak{R}[P_F]]_n, (M \otimes_A F)^n)$$

Consequently, the composite of $(a_M \otimes 1 - 1 \otimes a_F)$ with $\operatorname{Hom}_{\mathbb{Z}S_n}(j, p^n) \circ \mathfrak{f}[M \otimes F]$ is *nullhomotopic*. This isn't quite what we need in order to apply lemma 2.1 on page 86. We want a null-homotopy of the restriction of $\operatorname{Hom}_{\mathbb{Z}S_n}(j, p^n) \circ \mathfrak{f}[M \otimes F]$ to the *kernel* of p — i.e. we want to *transport* this nullhomotopy of $\operatorname{Hom}_{\mathbb{Z}S_n}(j, p^n) \circ \mathfrak{f}[M \otimes F] \circ (a_M \otimes 1 - 1 \otimes a_F)$ to $M \otimes F$. Assumption 5 implies that this is possible: it says that the nullhomotopy vanishes on the kernel of $a_M \otimes 1 - 1 \otimes a_F: M \otimes A \otimes F \rightarrow M \otimes F$ so that it can be "pushed forward" to $M \otimes F$. The map $z: M \otimes F \rightarrow M \otimes A \otimes F$ accomplishes this "pushing forward".

Since the resulting map (after "pushing forward" the null homotopy) is well-defined over all of $M \otimes F$, we can subtract $\partial(\operatorname{Hom}_{\mathbb{Z}S_n}(w_1, p^n) \circ V_1 \circ \Phi_1 \circ \mathfrak{f}[F] \circ z - \operatorname{Hom}_{\mathbb{Z}S_n}(w_2, p^n) \circ V_2 \circ \mathfrak{f}[M] \otimes \Phi_2 \circ z)$ from $\operatorname{Hom}_{\mathbb{Z}S_n}(j, p^n) \circ \mathfrak{f}[M \otimes F]$ to get a map whose restriction to the kernel of p vanishes. This completes the proof. $\square$

The following result makes the procedure of taking a tensor product of m-modules over an m-Hopf algebra (see 3.13 on page 26 for a definition) well-defined in the category of m-coalgebras. Note that something along these lines must be done because the action of an m-Hopf algebra on an m-module only preserves m-structures up to a morphism of m-coalgebras.

COROLLARY 2.4. *Let M be a strict free right m-module over an m-Hopf algebra A, and let F be a left m-module over A. Then the tensor product $M \otimes_A F$ can be equipped with an m-structure*

$$\mathfrak{f}_x: M \otimes_A F \rightarrow \operatorname{Hom}_{\mathbb{Z}S_n}([\mathfrak{R}[M] \otimes^{\mathfrak{R}[A]} \mathfrak{R}[P_F]]_n (M \otimes_A F)^n)$$

such that $p: M \otimes F \rightarrow M \otimes_A F$, the projection that identifies the right action of A on M with the left action of A on F, preserves m-structures up to a chain-homotopy. Let $(\iota_F, \varphi_F, r_F, g_F): A \otimes F \rightarrow P_F \rightarrow F$ denote the structure morphism of F. Then this m-structure is given by:

$$\operatorname{Hom}_{\mathbb{Z}S_n}(j, p^n) \circ \mathfrak{f}[M \otimes F] \circ t + \partial\{\operatorname{Hom}_{\mathbb{Z}S_n}(w_2, p^n) \circ V_2 \circ \mathfrak{f}[B] \otimes \Phi\} \circ t: M \otimes_A F$$
$$\rightarrow \operatorname{Hom}_{\mathbb{Z}S_n}([\mathfrak{R}[M] \otimes^{\mathfrak{R}[A]} \mathfrak{R}[P_F]]_n, (M \otimes AF)^n)$$

where:

1. *$t: M \otimes_A F \rightarrow M \otimes F$ is a right-inverse to p;*

2. $\mathfrak{f}[B] = \mathfrak{f}[M]|B \otimes 1 \colon B \to \mathrm{Hom}_{\mathbb{Z} S_n}(\mathfrak{R}[M]_n, M^n)$, *where* $M = B \otimes A$ *(as a graded right A-module), and B is a $\mathbb{Z}$-free module generated by a free basis for M over A;*

3. *V is a shuffle map that sends* $(M^n \otimes F^n)$ *to* $(M \otimes F)^n$;

4. $F = r_F^n \circ \mathfrak{R}[P_F] \circ \varphi_F \circ \iota_F \colon A \otimes F \to \mathrm{Hom}_{\mathbb{Z} S_n}(\mathfrak{R}[P_F]_n, F^n)$, *the chain-homotopy between* $\mathfrak{f}[F] \circ a_F$ *and* $a_F^n \circ \mathfrak{f}[A \otimes F]$, *where* $a_F \colon A \otimes F \to F$ *represents the action of A on F.*

5. *The ring A has no zero-divisors;*

6. $V_2 \colon \mathrm{Hom}_{\mathbb{Z} S_n}(\mathfrak{R}[M]_n, M^n) \otimes \mathrm{Hom}_{\mathbb{Z} S_n}(\mathfrak{R}[P_F]_n, F^n) \to \mathrm{Hom}_{\mathbb{Z} S_n}(\mathfrak{R}[M]_n \otimes \mathfrak{R}[P_F]_n, (M \otimes F)^n)$ *is the shuffle map.*

7. $w_2 \colon [\mathfrak{R}[P_M] \otimes^{\mathfrak{R}[A]} \mathfrak{R}[P_F]] \to \mathfrak{R}[M] \otimes \mathfrak{R}[P_F]$ *is the natural map.*

REMARK. 2.4.1. In the statement $\cdots \circ \mathfrak{f}[B] \otimes F\} \circ t$ we regard $M \otimes F$ as $B \otimes A \otimes F$. The chain-homotopy Φ is applied to the factors $A \otimes F$.

PROOF. This is a direct consequence of 2.3 on page 87. That lemma implies that the given m-structure has the property that the map p preserves m-structures up to a chain-homotopy. Our hypotheses imply that Φ_1 (in 2.3) vanishes and Φ (*here*) is equal to Φ_2 in that result. The assumption that M is a right free A-module and that A has no zero-divisors implies that the kernel of $a_M \otimes 1 - 1 \otimes a_F \colon M \otimes A \otimes F \to M \otimes F$ is precisely $M \otimes 1 \otimes A$. The last line in 3.15 on page 27 implies that requirement 5 of 2.3 on page 87 is satisfied.

The definition of $\mathfrak{f}[B]$ is derived from requirement 2 in 2.3, that there exist $z \colon M \otimes F \to M \otimes A \otimes F$, that is a right-inverse to $a_M \otimes 1 - 1 \otimes a_F \colon M \otimes A \otimes F \to M \otimes F$. We will calculate the image of $a_M \otimes 1 - 1 \otimes a_F$: it is equal to the kernel of $M \otimes F \to M \otimes_A F$. Since $M = B \otimes A$, $M \otimes_A F = B \otimes F$, and the kernel of this map is precisely $B \otimes \ker(a_F)$. The composite of z with $a_M \otimes 1 - 1 \otimes a_F$ must essentially define a retraction onto this submodule of $M \otimes F$. We can accomplish this by defining z to send $m \otimes x$ to $m \otimes x - w_1(m) \otimes 1 \otimes a_F(w_2(m) \otimes x)$, where w_1 is projection $M \to M \otimes_A \mathbb{Z} = B$, and $w_2(m)$ is such that $m = w_1(m) \otimes w_2(m)$, whenever m is of the form $b \otimes a$. Since $\Phi(1 \otimes a_F(w_2(m) \otimes x)) = 0$ (due to the last line in 3.15 on page 27), we can ignore this term and define z to be the map defined on elements of $M \otimes F$ of the form $m \otimes f$ (where m is of the form $b \otimes a$) via $b \otimes 1 \otimes a \otimes x - w_1(m) \otimes 1 \otimes 1 \otimes a_F(w_2(m) \otimes x)$. The composite of this with $\cdots \circ \mathfrak{f}[M] \otimes \Phi_2 \circ z$ is $\cdots \circ \mathfrak{f}[B] \otimes \Phi$, as stated in this result. □

COROLLARY 2.5. *Let C be a weakly coherent m-coalgebra, and let F be a left m-module over* $\mathcal{F}(C)$. *Then the twisted tensor product* $C \otimes_x F$ *can be equipped with an m-structure* $f_x \colon C \otimes_x F \to \mathrm{Hom}_{\mathbb{Z} S_n}([\mathcal{P}(\mathfrak{R}[C], n) \otimes^{\mathcal{F}(\mathfrak{R}[C], n)} \mathfrak{R}[P_F]]_n, (C \otimes_x F)^n)$, *such that* $p \colon C \otimes_\ell \mathcal{F}(C) \otimes F \to C \otimes_x F$, *the projection that identifies the right action of* $\mathcal{F}(C)$ *on* $C \otimes_\ell \mathcal{F}(C)$ *with its left action on F, preserves m-structures up to a chain-homotopy. Let* $(\iota_F, \varphi_F, r_F, g_F) \colon \mathcal{F}(C) \otimes F \to P_F \to F$ *denote the structure morphism of F. Then this m-structure is given by:*

$$\mathrm{Hom}_{\mathbb{Z} S_n}(w_2, p^n) \circ V_2 \circ \mathfrak{f}[C] \otimes \mathfrak{f}[F] + \mathfrak{f}[C] \otimes \Phi \circ (1 \otimes \downarrow \otimes 1) \circ (\Delta_C \otimes 1)$$

$$\colon C \otimes_x F \to \mathrm{Hom}_{\mathbb{Z} S_n}([\mathcal{P}(\mathfrak{R}[C], n) \otimes^{\mathcal{F}(\mathfrak{R}[C], n)} \mathfrak{R}[P_F]]_n, (C \otimes_x F)^n)$$

where:

1. $\mathfrak{f}[C] = \mathcal{A}(f_n)|C \otimes 1 \colon C \to \mathrm{Hom}_{\mathbb{Z} S_n}(\mathcal{P}(\mathfrak{R}[C], n), (C \otimes_\ell \mathcal{F}(C))^n)$;

2. $\Phi = r_F^n \circ \mathfrak{f}[P_F] \circ \varphi_F \circ \iota_F \colon \mathcal{F}(C) \otimes F \to \mathrm{Hom}_{\mathbb{Z}S_n}(\mathfrak{R}[P_F]_n, F^n)$, the chain-homotopy between $\mathfrak{f}[F] \circ a_F$ and $a_F^n \circ \mathfrak{f}[\mathcal{F}(C) \otimes F]$, where $a_F \colon \mathcal{F}(C) \otimes F \to F$ represents the action of $\mathcal{F}(C)$ on F.

$$V_2 \colon \mathrm{Hom}_{\mathbb{Z}S_n}(\mathcal{P}(\mathfrak{R}[C], n), (C \otimes_\ell \mathcal{F}(C))^n) \otimes \mathrm{Hom}_{\mathbb{Z}S_n}(\mathfrak{R}[P_F]_n, F^n)$$
$$\to \mathrm{Hom}_{\mathbb{Z}S_n}(\mathcal{P}(\mathfrak{R}[C], n) \otimes \mathfrak{R}[P_F], (C \otimes_\ell \mathcal{F}(C) \otimes F)^n)$$

is the shuffle map. $w_2 \colon [\mathcal{P}(\mathfrak{R}[C], n) \otimes^{\mathcal{F}(\mathfrak{R}[C], n)} \mathfrak{R}[P_F]] \to \mathcal{P}(\mathfrak{R}[C], n) \otimes \mathfrak{R}[P_F]$ is the natural map.

REMARK. 2.5.1. $\mathcal{P}(\mathfrak{R}[C], n) = \hat{Y}_n(\mathfrak{R}[C]) \bigstar_{\{\ell, \hat{\mu}_i\}} \mathcal{F}(\mathfrak{R}[C], n)$ is the coordinate coalgebra of $C \otimes_\ell \mathcal{F}(C)$—see 1.15 on page 82. The cotensor product $\mathcal{P}(\mathfrak{R}[C], n) \otimes^{\mathcal{F}(\mathfrak{R}[C], n)} \mathfrak{R}[P_F]$ is equal to $\hat{Y}_n(\mathfrak{R}[C]) \bigstar_{\{\ell, \hat{\mu}_i\}} \mathcal{F}(\mathfrak{R}[C], n)$.

PROOF. This is a straightforward application of 2.4 on page 90. The ring $\mathcal{F}(C)$ is free, so it has no zero-divisors. The map t clearly has the desired properties. The formula for the m-structure is derived from that in 2.4 using the fact that $* \otimes \Phi \circ t = 0$:

$$\partial \{ \mathrm{Hom}_{\mathbb{Z}S_n}(w_2, p^n) \circ V_2 \circ \mathfrak{f}[B] \otimes \Phi \} \circ t$$
$$= \partial \circ \{ \mathrm{Hom}_{\mathbb{Z}S_n}(w_2, p^n) \circ V_2 \circ \mathfrak{f}[B] \otimes F \} \circ t + \{ \mathrm{Hom}_{\mathbb{Z}S_n}(w_2, p^n) \circ V_2 \circ \mathfrak{f}[B] \otimes F \} \circ \partial \circ t$$
$$= \{ \mathrm{Hom}_{\mathbb{Z}S_n}(w_2, p^n) \circ V_2 \circ \mathfrak{f}[B] \otimes F \} \circ \partial \circ t (\text{since } * \otimes \Phi \circ t = 0)$$
$$= \{ \mathrm{Hom}_{\mathbb{Z}S_n}(w_2, p^n) \circ V_2 \circ \mathfrak{f}[C] \otimes \Phi \} \circ (1 \otimes \downarrow \otimes 1) \circ (\Delta_C \otimes 1)$$

Here, the untwisted portion of the differential of $C \otimes_\ell \mathcal{F}(C)$ is annihilated by $* \otimes \Phi$ so we are left with the twisted portion. $\square$

COROLLARY 2.6. Let C be a weakly coherent m-coalgebra, let H be a coherent m-Hopf algebra, and let $g \colon C \to \bar{\mathcal{B}}(H)$ be a strict morphism of weakly coherent m-coalgebras. Then the twisted tensor product $C \otimes_\xi H$, where $\xi = f(x) \circ \mathcal{F}(g) \circ x$, can be equipped with an m-structure $f_\xi \colon C \otimes_\xi H \to \mathrm{Hom}_{\mathbb{Z}S_n}(\mathcal{P}(\mathfrak{R}[C], n) \otimes^{\mathcal{F}(\mathrm{R}(S_n))} \mathrm{R}(S_n), (C \otimes_\xi H)^n)$, such that $p \colon C \otimes_\ell \mathcal{F}(C) \otimes H \to C \otimes_\xi H$, the projection that identifies the right action of $\mathcal{F}(C)$ on $C \otimes_\ell \mathcal{F}(C)$ with its left action on H, preserves m-structures up to a chain-homotopy. This m-structure is given by:

$$\mathrm{Hom}_{\mathbb{Z}S_n}(w_2, p^n) \circ V_2 \circ \mathfrak{f}[C] \otimes \mathfrak{f}[H]$$
$$\colon C \otimes_\xi H \to \mathrm{Hom}_{\mathbb{Z}S_n}(\mathcal{P}(\mathfrak{R}[C], n) \otimes^{\mathcal{F}(\mathrm{R}(S_n))} \mathrm{R}(S_n), (C \otimes_\xi H)^n)$$

where: $\mathfrak{f}[C] = \mathcal{A}(f_n) | C \otimes 1 \colon C \to \mathrm{Hom}_{\mathbb{Z}S_n}(\mathcal{P}(\mathfrak{R}[C], n), (C \otimes_\ell \mathcal{F}(C))^n)$;

$$V_2 \colon \mathrm{Hom}_{\mathbb{Z}S_n}(\mathcal{P}(\mathfrak{R}[C], n), (C \otimes_\ell \mathcal{F}(C))^n) \otimes \mathrm{Hom}_{\mathbb{Z}S_n}(\mathrm{R}(S_n), H^n)$$
$$\to \mathrm{Hom}_{\mathbb{Z}S_n} \mathcal{P}(\mathfrak{R}[C], n) \otimes \mathrm{R}(S_n), (C \otimes_\ell \mathcal{F}(C))^n)$$

is the shuffle map.
$w_2 \colon [\mathcal{P}(\mathfrak{R}[C], n) \otimes^{\mathcal{F}(\mathrm{R}(S_n), n)} \mathrm{R}(S_n)] \to \mathcal{P}(\mathfrak{R}[C], n) \otimes \mathrm{R}(S_n)$ is the natural map.

Remarks.   2.6.1. The $\mathcal{F}(\mathrm{R}(S_n), n)$-comodule structure of $\mathrm{R}(S_n)$ is defined by forming the composite $(h_n \otimes 1) \circ \Delta_R$, where $\Delta_R \colon \mathrm{R}(S_n) \to \mathrm{R}(S_n) \otimes \mathrm{R}(S_n)$ is the coproduct and $h_n \colon \mathrm{R}(S_n) \to \mathcal{F}(\mathrm{R}(S_n), n)$ is the coalgebra morphism defined in 2.26 on page 55. The cotensor product $\mathcal{P}(\mathfrak{R}[C], n) \otimes^{\mathcal{F}(\mathrm{R}(S_n))} \mathrm{R}(S_n)$ is equal to $\hat{Y}_n(\mathfrak{R}[C]) \bigstar_{h_n \circ \{\ell, \hat{\mu}_i\}} \mathrm{R}(S_n)$.

2.6.2. We will refer to the m-coalgebra structure defined here as the canonical m-coalgebra structure on $C \otimes_\xi H$.

PROOF. First, we claim that the m-coalgebra structure described above must be weakly-coherent. This is because it is a tensor product of a coherent m-coalgebra structure of H and the restriction of a weakly-coherent m-coalgebra structure of $C \otimes_\ell \mathcal{F}(C)$. It only remains to show that it has the stated form — i.e., that $f_\xi \colon C \otimes_\xi H \to \mathrm{Hom}_{\mathbb{Z} S_n}(\mathcal{P}(\mathfrak{R}[C], n) \otimes^{\mathcal{F}(\mathrm{R}(S_n))} \mathrm{R}(S_n), (C \otimes_\xi H)^n)$ is a chain-map. This follows from 2.5 on page 91 and the explicit description of the contraction $(f(x), \mathcal{I}, \phi_\mathcal{I}) \colon \mathcal{F}(\bar{\mathcal{B}}(C)) \to C$ given in 3.11 on page 65. We use this contraction to construct the left-action of $\mathcal{F}(\bar{\mathcal{B}}(C))$ on C, and use the formula in 2.5 to give the m-structure map $\mathrm{Hom}_{\mathbb{Z} S_n}(w_2, p_n) \circ V_2 \circ \{f[C] \otimes f[H] + f[C] \otimes \cdots \otimes 1 \circ (1 \otimes \ell \otimes 1) \circ (\Delta_C \otimes 1)\}$ for $C \otimes_\xi H$. But the description of $\phi_\mathcal{I}$ given in 3.11 on page 65 implies that the composite $1 \otimes \circ (1 \otimes \ell) \circ \Delta_C$ is identically zero. $\square$

3. Geometricity of the m-structure on a twisted tensor product

In this section we will apply the results of this paper to computing m-structures on the chain complexes of total spaces of fibrations. These m-structures are geometrically relevant in the sense that they are homotopy equivalent to the ones that result from applying the functor $\mathcal{C}(*)$ (defined on page 29) to the total space of the fibration. They can, consequently, be used to compute the cohomology ring structure and the cohomology operations of the total spaces of fibrations.

LEMMA 3.1. *Let $X \times_\xi F$ be a twisted cartesian product, where F is a left topological module over ΩX, and X is a pointed simply-connected space. Then there exists a twisting cochain $\bar{\xi} \colon \mathcal{C}(X) \to \mathcal{C}(\Omega X)$ with associated twisted tensor product, and a natural contraction $(\varphi_\xi, f_\xi, g_\xi) \colon \mathcal{C}(X \times_\xi F) \to \mathcal{C}(X) \otimes_{\bar{\xi}} \mathcal{C}(F)$. The maps in this contraction (including the chain-homotopy) preserve the action of $\mathcal{C}(\Omega X)$ on $\mathcal{C}(F)$.*

REMARK. 3.1.1. The fact that the maps in this contraction preserve the action of ΩX on F implies that we can use them to "transport" the canonical m-coalgebra structure of $\mathcal{C}(X \times_\xi F)$ to to get an m-structure (usually not coherent) $\bar{f}_\xi \colon \mathcal{C}(X) \otimes_{\bar{\xi}} \mathcal{C}(F) \to \mathrm{Hom}_{\mathbb{Z} S_n}(\mathrm{R}(S_n), (\mathcal{C}(X) \otimes_{\bar{\xi}} \mathcal{C}(F))^n)$. We will call an m-structure on a twisted tensor product that has been transported from a twisted cartesian product in this way a *geometric* m-coalgebra structure. These geometric m-structures fit into a homotopy-commutative diagram:

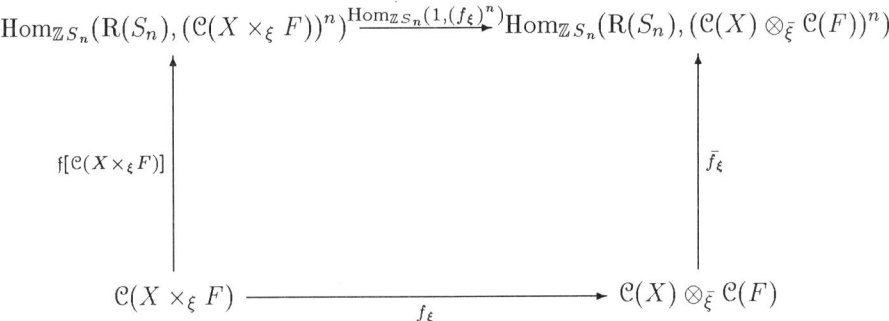

$$\operatorname{Hom}_{\mathbb{Z}S_n}(\mathrm{R}(S_n),(\mathcal{C}(X\times_\xi F))^n)\xrightarrow{\operatorname{Hom}_{\mathbb{Z}S_n}(1,(f_\xi)^n)}\operatorname{Hom}_{\mathbb{Z}S_n}(\mathrm{R}(S_n),(\mathcal{C}(X)\otimes_{\bar\xi}\mathcal{C}(F))^n)$$

$$f[\mathcal{C}(X\times_\xi F)]\qquad\qquad\qquad\qquad\qquad\bar{f}_\xi$$

$$\mathcal{C}(X\times_\xi F)\xrightarrow{\ \ f_\xi\ \ }\mathcal{C}(X)\otimes_{\bar\xi}\mathcal{C}(F)$$

Let X and F be simplicial complexes and $a\colon\Omega X\times F\to F$ be a simplicial mapping defining an action of ΩX on F. Then the induced chain-map $\bar{a}=g\circ a\colon\mathcal{C}(\Omega X)\otimes\mathcal{C}(F)\to\mathcal{C}(F)$ (where $g\colon\mathcal{C}(\Omega X)\otimes\mathcal{C}(F)\to\mathcal{C}(\Omega X\times F)$ is the map from the Eilenberg-Zilber theorem) is a morphism of m-coalgebras, by 4.5 on page 31. We can make use of the m-module structure of $\mathcal{C}(X)\otimes_\alpha\mathcal{C}(\Omega X)$ over $\mathcal{C}(\Omega X)$ (see 3.15 on page 27) to define an m-structure on $\mathcal{C}(X)\otimes_\xi\mathcal{C}(F)$ by forming the tensor product $\mathcal{C}(X)\otimes_\alpha\mathcal{C}(\Omega X)\otimes_{\mathcal{C}(\Omega X)}\mathcal{C}(F)$ — and using 2.3 on page 87.

PROPOSITION 3.2. *Given an m-coalgebra structure on $\mathcal{C}(X)\otimes_\alpha\mathcal{C}(\Omega X)$ that makes it an m-module over $\mathcal{C}(\Omega X)$ there exists a unique m-coalgebra structure, f', on that makes the diagram in figure 4.4.2 (on page 95) commute up to a homotopy.*

Corollary 2.4 on page 90 gives an explicit formula for this m-structure.

We will now derive equations for m-coalgebra structures homotopic to these (as maps of $\mathcal{C}(X)\otimes_{\bar\xi}\mathcal{C}(F)$) to $\operatorname{Hom}_{\mathbb{Z}S_n}(\mathrm{R}(S_n),(\mathcal{C}(X)\otimes_{\bar\xi}\mathcal{C}(F))^n)$, for all $n\geq 2$ — that will be expressible in terms of homotopy theoretic invariants of X and F. Since these new structures are homotopic to the geometric ones they may also be regarded as geometric.

Note that $\mathcal{C}(X)\otimes_\alpha\mathcal{C}(\Omega X)$ can be regarded as a free $\mathcal{C}(\Omega X)$-module. This implies that we can define an m-structure on $\mathcal{C}(X)\otimes_\alpha\mathcal{C}(\Omega X)$ by defining it to be free over $\mathcal{C}(\Omega X)$. This amounts to specifying its values on basis elements (essentially basis $\mathbb{Z}$-elements of $\mathcal{C}(X)\otimes 1$) and extending the structure to all of $\mathcal{C}(X)\otimes_\alpha\mathcal{C}(\Omega X)$ by requiring it to be a right m-module over $\mathcal{C}(\Omega X)$. This implies that:

PROPOSITION 3.3. *We can define an m-structure $\hat{f}_{\mathcal{C}(X)\otimes_\alpha\mathcal{C}(\Omega X)}\colon\mathcal{C}(X)\otimes_\alpha\mathcal{C}(\Omega X)\to$ $\operatorname{Hom}_{\mathbb{Z}S_n}(\mathrm{R}(S_n),(\mathcal{C}(X)\otimes_\alpha\mathcal{C}(\Omega X))^n)$ via 2.4 on page 90 and $\mathcal{C}(X)\otimes_\alpha\mathcal{C}(\Omega X)=$ $\mathcal{C}(X)\otimes_\ell\mathcal{F}(\mathcal{C}(X))\otimes_{\mathcal{F}(\mathcal{C}(X))}\mathcal{C}(\Omega X)$. Furthermore, the following diagram is chain-homotopy*

Figure 4.4.2.

commutative:

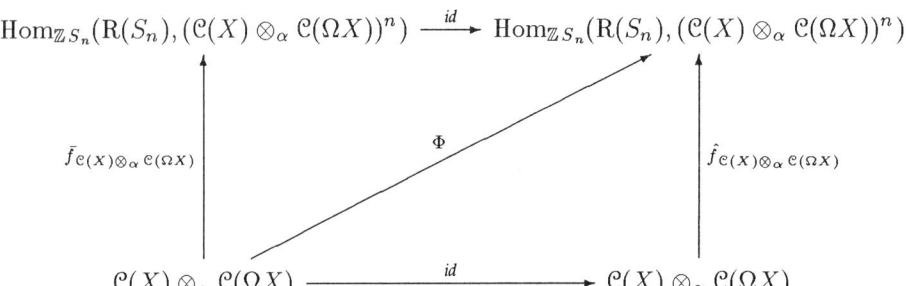

Remarks. 3.3.1. This implies that the m-structure $\hat{f}_{\mathcal{C}(X)\otimes_\alpha \mathcal{C}(\Omega X)}: \mathcal{C}(X) \otimes_\alpha \mathcal{C}(\Omega X) \to \operatorname{Hom}_{\mathbb{Z}S_n}(\mathrm{R}(S_n),(\mathcal{C}(X) \otimes_\alpha \mathcal{C}(\Omega X))^n)$ is geometric, in the sense defined above. This is significant because it is based upon $\mathcal{C}(X) \otimes_\ell \mathcal{F}(\mathcal{C}(X))$, which is a strict m-module over $\mathcal{F}(\mathcal{C}(X))$ and has a computable m-structure — given in 2.30 on page 57. Naturality of this construct implies that a similar m-structure can be defined on (in 3.4) that is also computable and geometric.

3.3.2. Here we make use of the natural chain-homotopy equivalence of m-Hopf algebras $F: \mathcal{F}(\mathcal{C}(X)) \to \mathcal{C}(\Omega X)$ defined in 4.2 on page 66.

3.3.3. Corollary 2.4 on page 90 can be used to get an explicit formula for this m-structure.

3.3.4. The fact that this diagram commutes up to a chain-homotopy follows from the fact that the chain-complexes in question are *all acyclic.*

PROOF. This is a consequence of 2.5 on page 91. □

COROLLARY 3.4. *We can define an m-structure,* $\hat{f}_{\mathcal{C}(X)\otimes_{\bar{\xi}}\mathcal{C}(F)}: \mathcal{C}(X) \otimes_{\bar{\xi}} \mathcal{C}(F) \to \operatorname{Hom}_{\mathbb{Z}S_n}(\mathrm{R}(S_n),(\mathcal{C}(X) \otimes_{\bar{\xi}} \mathcal{C}(F))^n)$ *via* 2.4 *on page* 90 *and* $\mathcal{C}(X) \otimes_{\bar{\xi}} \mathcal{C}(F) = (\mathcal{C}(X) \otimes_\ell \mathcal{F}(\mathcal{C}(X))) \otimes_{\mathcal{F}(\mathcal{C}(X))} \mathcal{C}(F),$ *where the left-action of* $\mathcal{F}(\mathcal{C}(X))$ *is defined by composition with* $F: \mathcal{F}(\mathcal{C}(X)) \to \mathcal{C}(\Omega X)$ *(the map defined in 4.2 on page 66). In addition, this m-structure is homotopic to that of* $(\mathcal{C}(X) \otimes_\ell \mathcal{F}(\mathcal{C}(X))) \otimes_{\mathcal{F}(\mathcal{C}(X))} \mathcal{C}(\Omega X) \otimes_{\mathcal{C}(\Omega X)} \mathcal{C}(F),$ *which is homotopic to the geometric m-structure on defined in remark 3.1.1 on page 93.*

COROLLARY 3.5. *Let* $f: X \to Y$ *be a map of pointed simply connected spaces. The homotopy-fiber of this map is homotopy-equivalent (as an m-coalgebra) to the weakly-coherent m-coalgebra* $(\mathcal{C}(X)_\ell \mathcal{F}(\mathcal{C}(X))) \otimes_{\mathcal{F}(\mathcal{C}(X))} \mathcal{F}(\mathcal{C}(Y)) = \mathcal{C}(X) \otimes_{\mathcal{F}(f)\circ\ell} \mathcal{F}(\mathcal{C}(Y)).$

The chain-maps $u_n = c_{n,m} \otimes h_n \circ \Delta_R: \mathrm{R}(S_n) \to \hat{Y}_n(\mathfrak{R})\bigstar_{\{\ell,\hat{\mu}_i\}}\mathcal{F}(\mathfrak{R},n)$, defined in 1.13 on page 81, can be used to explicitly compute the m-structure on twisted tensor products. We form the composite of the structure maps

$$\mathfrak{f}[\mathcal{C}(X) \otimes_{\mathcal{F}(f)\circ\ell} \mathcal{F}(\mathcal{C}(Y))]_n: \mathcal{C}(X) \otimes_{\mathcal{F}(f)\circ\ell} \mathcal{F}(\mathcal{C}(Y))$$
$$\to \operatorname{Hom}_{\mathbb{Z}S_n}(\hat{Y}_n(\mathfrak{R})\bigstar_{\{\ell,\hat{\mu}_i\}}\mathcal{F}(\mathfrak{R},n),(\mathcal{C}(X) \otimes_{\mathcal{F}(f)\circ\ell} \mathcal{F}(\mathcal{C}(Y)))^n)$$
$$\to \operatorname{Hom}_{\mathbb{Z}S_n}(\mathrm{R}(S_n),(\mathcal{C}(X) \otimes_{\mathcal{F}(f)\circ\ell} \mathcal{F}(\mathcal{C}(Y)))^n)$$

where the map on the right is $\operatorname{Hom}_{\mathbb{Z} S_n}(u_n, 1)$. This composite can be used to compute higher-coproducts for $\mathcal{C}(X) \otimes_{\mathcal{F}(f) \circ \ell} \mathcal{F}(\mathcal{C}(Y))$.

Appendices

A. Proof of 2.17 (on page 17)

It is easy to see that the maps in question are *chain-maps:* they are composites of maps that have been proved to be chain-maps. This implies that the composition operations have the correct behavior with respect to the differential (i.e., the definition of the differential on the tensor product $R(S_m) \otimes R(S_n)$ implies the statement). It is necessary to prove the composition-properties — i.e., statements

- $(S_1 \circ_i S_2) \circ_j S_3 = S_1 \circ_{i+j-1} (S_2 \circ_j S_3)$;
- if $j < i$ then $S_1 \circ_{i+\mathrm{rank}(S_2)-1} (S_2 \circ_j S_3) = (-1)^{\dim(S_2) \cdot \dim()} S_2 \circ_j (S_1 \circ_i S_3)$

Throughout this proof we will use the following notation:

DEFINITION A.1. $\mathrm{T}_{\underbrace{1,\ldots,m,\ldots,1}_{i^{\text{th}} \text{ position}}} = \hat{\mathrm{T}}(i,m)$, and $\mathfrak{T}_{\underbrace{1,\ldots,m,\ldots,1}_{i^{\text{th}} \text{ position}}} = \hat{\mathfrak{T}}(i,m)$.

PROPOSITION A.2. *Let i, j, m, n, N be integers ≥ 1, and assume $i, j, m, n \leq N$. Then:*
$\hat{\mathrm{T}}(i,m) \circ \hat{\mathrm{T}}(j,n) =$

1. $\mathrm{T}_{\underbrace{1,\ldots,n,\ldots,m,\ldots,1}_{\substack{j^{\text{th}} \text{ and } i-n^{\text{th}} \\ \text{positions,} \\ \text{respectively}}}}$ *if $i > n + j$;*

2. $\hat{\mathrm{T}}(j, m + n - 1)$ *if $j \leq i \leq n + j$;*

3. $\mathrm{T}_{\underbrace{1,\ldots,m,\ldots,n,\ldots,1}_{\substack{i^{\text{th}} \text{ and } j^{\text{th}} \\ \text{positions,} \\ \text{respectively}}}}$ *if $i < j$.*

   We will say that an element $[g_1 | \ldots | g_k] \in R(S_N)$ stabilizes a range of numbers $\{a, \ldots, b\}$ if all of the permutations $g_1, \ldots, g_k$ map this range into itself. Then
   $\hat{\mathfrak{T}}(i,m) \circ \hat{\mathfrak{T}}(j,n)(\sigma) =$

4. $\mathfrak{T}_{\underbrace{1,\ldots,n,\ldots,m,\ldots,1}_{\substack{j^{\text{th}} \text{ and } i-n^{\text{th}} \\ \text{positions,} \\ \text{respectively}}}}(\sigma)$ *if $i > n + j$, and σ stabilizes $\{1, \ldots, n + j\}$;*

5. $\hat{\mathfrak{T}}(j, m + n - 1)(\sigma)$ *if $j \leq i \leq n + j$, and σ stabilizes $\{i, \ldots, n + j\}$;*

6. $\mathfrak{T}_{\underbrace{1,\ldots,m,\ldots,n,\ldots,1}}(\sigma)$ if $i < j$, and σ stabilizes $\{1,\ldots,j-1\}$.

$\quad\quad\quad\; i^{\text{th}}$ and j^{th}
$\quad\quad\quad\quad$ positions,
$\quad\quad\quad\quad$ respectively

In addition:

7. $R(\mathcal{S}_{i+n-1}) = \hat{\mathfrak{T}}(j,n) \circ R(\mathcal{S}_i)$, if $j < i$.

Remarks.   A.2.1. The first three statements follow immediately from the definition of the T-maps in 2.7. The second three statements follow from 2.10 and the fact that stabilizing a given range of integers in $\{1,\ldots,N\}$ is equivalent to stabilizing its complement.

A.2.2. Statement 7 is clear from upon inspection of a permutation-representation of $R(\mathcal{S}_i)(\sigma)$, where $\sigma \in S_n$:

$$\begin{pmatrix} 1 & \cdots & j & \cdots & i & \cdots & i\mid n & 1 \\ 1 & \cdots & j & \cdots & \sigma(1) & \cdots & \sigma(n) & \end{pmatrix}$$

The effect of $\hat{\mathfrak{T}}(j,n)$ is just to add $n-1$ more indices to column j.

PROPOSITION A.3. *Let α, i, j, n, and m be integers ≥ 1 and with $i \leq n$, and $j \leq \alpha \leq m+n-1$. Then the following identity holds in $R(S_{n+m-1})$:*

$$\hat{\mathfrak{T}}(\alpha,m) \circ \circledast \left\{ R(\mathcal{S}_{j-1}) \otimes \hat{\mathfrak{T}}(j,n) \right\} = \circledast \left\{ \hat{\mathfrak{T}}(\alpha,m) \circ R(\mathcal{S}_{j-1}) \otimes \hat{\mathfrak{T}}(j,m+n-1) \right\}$$

REMARK. A.3.1. The result essentially implies that $\hat{\mathfrak{T}}(\alpha,m)$ can be permuted with $\circledast$.

PROOF. We apply these expressions to $S_1 \otimes S_2 \in R(S_m) \otimes R(S_n)$ and perform induction on the dimension of S_2. The statement is clear in the case where the dimension of S_2 is zero. Now suppose $R(\mathcal{S}_{j-1})(S_1) = [a_1|\ldots|a_m]$, $S_2 = [b_1|\ldots|b_n]$, and the result is valid in the case where the dimension of S_2 is $n-1$. Now:

1. $\hat{\mathfrak{T}}(j,n)(S_2) = [\hat{T}(j,n)(b_1)|\hat{\mathfrak{T}}(b_1^{-1}j,n)(b_2|\ldots|b_n])$;

2.

(A.1) $R(\mathcal{S}_{j-1})(S_1) \circledast \hat{\mathfrak{T}}(j,n)(S_2)$

$\qquad = (-1)^m [z([z^{-1}a_1 z|\ldots|z^{-1}a_m z]) \circledast \hat{\mathfrak{T}}(b_1^{-1}j,n)(b_2|\ldots|b_n])$

$\qquad + (-1)^{m-1}[a_1|z([z^{-1}a_2 z|\ldots|z^{-1}a_m z]) \circledast \hat{\mathfrak{T}}(b_1^{-1}j,n)(b_2|\ldots|b_n])$

$\qquad\qquad\qquad + \cdots + [a_1|\ldots|a_m|z|\hat{\mathfrak{T}}(b_1^{-1}j,n)(b_2|\ldots|b_n])$

where $z = \hat{T}(j,n)(b_1)$, by 2.10.

Now we apply $\hat{\mathfrak{T}}(\alpha,m)$ to the second expression. We initially get

(A.2) $\hat{\mathfrak{T}}(\alpha,m) \circ R(\mathcal{S}_{j-1})(S_1) \circledast \hat{\mathfrak{T}}(j,n)(S_2)$

$\qquad = (-1)^m [z[z^{-1}a_1 z]|\ldots|z^{-1}a_m z]) \circledast (b_2|\ldots|b_n]$

$\qquad + (-1)^{m-1}[a_1|z([z^{-1}a_2 z|\ldots|z^{-1}a_m z]) \circledast (b_2|\ldots|b_n])$

$\qquad\qquad\qquad + \cdots + [a_1|\ldots|a_m|z|(b_2|\ldots|b_n])$

where, again, $z = \hat{T}(j, n)(b_1)$.

Now $\hat{T}(\alpha, m)z = \hat{T}(\alpha, m)\hat{T}(j, n)(b_1) = \hat{T}(j, m + n - 1)(b_1)$ by statement 2 of A.2.

CLAIM A.3.1. $z = \hat{T}(j, n)(b_1)$ *maps numbers in the sequence* $\{b_1^{-1}(j), \ldots, b_1^{-1}(j) + n - 1\}$ *to corresponding numbers in the sequence* $\{j, \ldots, j + n - 1\}$. *This is apparent upon inspecting the representation for* $\hat{T}(j, n)(b_1)$:

$$\begin{pmatrix} 1 & \ldots & t & \ldots & t + n - 1 & \ldots & n + b_1(t) - 1 \\ b_1(1) & \ldots & b_1(t) & \ldots & b_1(t) + n - 1 & \ldots & b_1(n) \end{pmatrix}$$

where $b_1(t) = j$.

CLAIM A.3.2.   1. $z^{-1}\{j, \ldots, j + n - 1\} \subseteq \{b_1^{-1}(j), \ldots, b_1^{-1}(j) + n - 1\}$;
     2. *for all* $u \geq 1$, $a_u\{j, \ldots, j + n - 1\} \subseteq \{j, \ldots, j + n - 1\}$;
     3. $z^{-1}a_u z = a_u' \in R(S_{b_1^{-1}(j)})\{1, \ldots, n\}$.

(A.3) $R(S_{j-1})(S_1) \circledast \hat{\mathfrak{T}}(j, n)(S_2) =$

$(-1)^m [\hat{T}(j, m + n - 1)(b_1)| \circledast \{[a_1'| \ldots |a_m']) \otimes \hat{\mathfrak{T}}(b_1^{-1}j, m + n - 1)(b_2| \ldots |b_n]\}$

$+ (-1)^{m-1} [\hat{T}(\alpha, m)a_1|\hat{T}(a_1^{-1}\alpha, m)z|\circledast([a_2'| \ldots |a_m']) \circledast \hat{\mathfrak{T}}(b_1^{-1}j, m+n-1)(b_2| \ldots |b_n])$

$+ \cdots + \hat{T}(\alpha, m)[a_1| \ldots |a_m|z|\hat{\mathfrak{T}}(b_1^{-1}j, m + n - 1)(b_2| \ldots |b_n])$

But this sum is precisely equal to $\circledast \left(\hat{\mathfrak{T}}(\alpha, m) \circ R(S_{j-1}) \otimes \hat{\mathfrak{T}}(j, m + n - 1) \right)$, *which proves the result.*

THEOREM A.4. *(This is 2.17 on page 17) The graded differential module,* $\mathfrak{S}$, *equipped with the following composition operations, constitutes a formal coalgebra (in the sense of 2.1 on page 11):*

$$\circ_i = \circledast(R(S_{i-1}) \otimes \underbrace{\mathfrak{T}_{1, \ldots, m, \ldots, 1}}_{i^{th} \, position}) : R(S_m) \otimes R(S_n) \rightarrow R(S_{m+n-1})$$

We show this by direct computation based on the following claims:

CLAIM A.4.1. $S_j \circ \hat{T}(i, m) = \hat{T}(i + j, m) \circ S_j$. *This is an immediate consequence of* 2.8 *on page 14.*

This implies the following corresponding statement:

CLAIM A.4.2. $R(S_j) \circ \hat{\mathfrak{T}}(i, m) = \hat{\mathfrak{T}}(i + j, m) \circ R(S_j)$.

Let $S_1 \in R(S_m)$, $S_2 \in R(S_n)$, and $S_3 \in R(S_t)$. The composite in statement A.4.1 above is:

(A.4) $(S_1 \circ_i S_2) \circ_j S_3$

$= \circledast \circ R(S_{j-1}) \otimes \hat{\mathfrak{T}}(j, m + n - 1) \circ (\circledast \otimes 1) \circ R(S_{i-1}) \otimes \hat{\mathfrak{T}}(i, m) \otimes 1(S_1 \otimes S_2 \otimes S_3)$

(A.5) $= \circledast \circ 1 \otimes \hat{\mathfrak{T}}(j, m + n - 1) \circ (\circledast \otimes 1)$

$$\circ \left(R(\mathcal{S}_{i+j-1}) \otimes R(\mathcal{S}_{j-1})\hat{\mathfrak{T}}(i, m) \otimes 1 \right) (S_1 \otimes S_2 \otimes S_3)$$

(by the fact that $\circledast$ commutes with $R(\mathcal{S}_{i-1})$

(A.6) $= \circledast \circ 1 \otimes \hat{\mathfrak{T}}(j, m + n - 1) \circ (\circledast \otimes 1)$

$$\circ \left(R(\mathcal{S}_{i+j-2}) \otimes \hat{\mathfrak{T}}(i + j - 1, m) \, R(\mathcal{S}_{j-1}) \otimes 1 \right) (S_1 \otimes S_2 \otimes S_3)$$

(by claim A.4.2 on page 101)

$$= \circledast \circ (\circledast \otimes 1) \circ \left(R(\mathcal{S}_{i+j-2}) \otimes \hat{\mathfrak{T}}(i + j - 1, m) \, R(\mathcal{S}_{j-1}) \otimes 1 \right) (S_1 \otimes S_2 \otimes S_3)$$

$$= \circledast \circ (1 \otimes \circledast) \circ \left(R(\mathcal{S}_{i+j-2}) \otimes \hat{\mathfrak{T}}(i + j - 1, m) \, R(\mathcal{S}_{j-1}) \otimes 1 \right) (S_1 \otimes S_2 \otimes S_3)$$

(by the associativity of $\circledast$)

$$= \circledast \circ R(\mathcal{S}_{i+j-2}) \otimes \hat{\mathfrak{T}}(i + j - 1, m) \circ \circledast \left(R(\mathcal{S}_{j-1}) \otimes \hat{\mathfrak{T}}(j, n) \right) (S_1 \otimes S_2 \otimes S_3)$$

(by A.3 with $\alpha = i + j - 1$) — we have replaced

$$\circledast \left(\hat{\mathfrak{T}}(i + j - 1, m) \, R(\mathcal{S}_{j-1}) \otimes \hat{\mathfrak{T}}(j, m + n - 1) \right)$$

in the equation, by

$$\hat{\mathfrak{T}}(i + j - 1, m) \circ \circledast \left(R(\mathcal{S}_{j-1}) \otimes \hat{\mathfrak{T}}(j, n) \right)$$

$= S_1 \circ_{i+j-1} S_2 \circ_j S_3$

Now we verify statement 2 of 2.1. We start with: $S_1 \circ_{i+\mathrm{rank}(S_2)-1} S_2 \circ_j S_3$

(A.7) $(S_1 \circ_i S_2) \circ_j S_3$

$$= \circledast \circ R(\mathcal{S}_{i+n-2}) \otimes \hat{\mathfrak{T}}(i + n - 1, m) \circ \circledast \circ \left\{ R(\mathcal{S}_{i-1}) \otimes \hat{\mathfrak{T}}(j, n) \right\} (S_1 \otimes S_2 \otimes S_3)$$

(A.8)

$$= \circledast \circ (1 \otimes \circledast) \circ R(\mathcal{S}_{i+n-2}) \otimes \hat{\mathfrak{T}}(i+n-1, m) \circ R(\mathcal{S}_{i-1}) \otimes \mathfrak{T}_{\underbrace{1,\dots,n,\dots,m,\dots,1}}(S_1 \otimes S_2 \otimes S_3)$$
$$\qquad\qquad\qquad\qquad\qquad\qquad j^{\text{th}} \text{ and } i^{\text{th}}$$
$$\qquad\qquad\qquad\qquad\qquad\qquad \text{positions,}$$
$$\qquad\qquad\qquad\qquad\qquad\qquad \text{respectively}$$

(by A.2)

(A.9)

$$= \circledast \circ (1 \otimes \circledast) \circ R(\mathcal{S}_{i+n-2}) \otimes \hat{\mathfrak{T}}(i+n-1, m) \circ R(\mathcal{S}_{i-1}) \otimes \mathfrak{T}_{\underbrace{1,\dots,n,\dots,m,\dots,1}}(S_1 \otimes S_2 \otimes S_3)$$
$$\qquad\qquad\qquad\qquad\qquad\qquad j^{\text{th}} \text{ and } i^{\text{th}}$$
$$\qquad\qquad\qquad\qquad\qquad\qquad \text{positions,}$$
$$\qquad\qquad\qquad\qquad\qquad\qquad \text{respectively}$$

(by associativity of $\circledast$)

Now note that the $\circledast$ in $\circledast \otimes 1$ can be regarded as an *ordinary* (i.e., untwisted) shuffle product —this is a consequence of:

1. $R(\mathcal{S}_{i+n-2})(S_1)$ has components that are permutations of the ordered set $\{i + n - 1, \ldots, i + n + m - 2\}$;
2. $R(\mathcal{S}_{j-1})(S_2)$ has components that are permutations of the ordered set $\{j, \ldots, j + n - 1\}$;
3. $j < i$. This implies that $\hat{\mathfrak{T}}(i + n - 1, m \circ R(\mathcal{S}_{j-1}))(S_2) = R(\mathcal{S}_{j-1})(S_2)$.

It follows that we can rewrite the expression as

(A.10)
$$= \circledast \circ (* \otimes 1) \circ R(\mathcal{S}_{i+n-2}) \otimes R(\mathcal{S}_{j-1}) \otimes \underbrace{\mathfrak{T}_{1,\ldots,n,\ldots,m,\ldots,1}}_{\substack{j^{\text{th}} \text{ and } i^{\text{th}} \\ \text{positions,} \\ \text{respectively}}} (S_1 \otimes S_2 \otimes S_3)$$

Now we interchange S_1 and S_2 —this multiplies the expression by $(-1)^{\dim(S_1)\dim(S_2)}$:

(A.11)
$$= (-1)^{nm} \circledast \circ (1 \otimes \circledast) \circ R(\mathcal{S}_{j-1}) \otimes R(\mathcal{S}_{i+n-2}) \otimes \underbrace{\mathfrak{T}_{1,\ldots,n,\ldots,m,\ldots,1}}_{\substack{j^{\text{th}} \text{ and } i^{\text{th}} \\ \text{positions,} \\ \text{respectively}}} (S_2 \otimes S_1 \otimes S_3)$$

(A.12) $$= (-1)^{nm} \circledast \circ (1 \otimes \circledast) \circ R(\mathcal{S}_{j-1}) \otimes \hat{\mathfrak{T}}(j, n) R(\mathcal{S}_{i-1}) \otimes$$
$$\hat{\mathfrak{T}}(j, n)) \circ \hat{\mathfrak{T}}(i, m)(S_2 \otimes S_1 \otimes S_3)$$

(by A.2, statements 6 and 7)

(A.13) $$= (-1)^{nm} \circledast \circ (1 \otimes \hat{\mathfrak{T}}(j, n) \circ \circledast) \circ R(\mathcal{S}_{j-1}) \otimes \hat{\mathfrak{T}}(j, n) R(\mathcal{S}_{i-1}) \otimes$$
$$\hat{\mathfrak{T}}(i, m))(S_2 \otimes S_1 \otimes S_3)$$

(by A.3) $= (-1)^{\dim(S_1)\dim(S_2)} S_2 \circ_j (S_1 \circ_i S_3)$ $\quad \square$

B. A morphism from $\bar{W}X$ to $\bar{\mathcal{B}}(X)$

We begin by giving a slightly nonstandard definition of the $\bar{W}$-construction due to Eilenberg and MacLane (see [7]):

A simplicial complex, C, will be called an R-complex if C_i has a well-defined ring structure for each i, and the product of two simplices is a simplex. We assume that if a group is acting on C all face and degeneracy maps commute with the action as well as the multiplication operation on simplices. A zero-dimensional simplex is assumed to be a unit in dimension zero, and all degeneracies of it are units in higher dimensions.

DEFINITION B.1. If C is an R-complex we define the $\bar{W}$-construction of C, denoted $\bar{W}(C)$, to be an R-complex as follows:

1. $F_0(C) = \mathbb{Z}$; This is generated by $1_{0,W}$, the unit.
2. $\bar{W}_{n+1}(C) = \bar{W}_n(C) \otimes C_n$;

3. $F_i(a \otimes b) = F_i a \otimes F_i b, 0 \le i < n-1$;
4. $F_{n-1}(a \otimes b) = a \cdot (1_{n-1,W} \otimes F_{n-1}b)$ (ring multiplication), where $a \in \bar{W}_{n-1}(C)$ and $b \in C_{n-1}$;
5. $F_n(a \otimes b) = \alpha(b)1_{n,W} \cdot a$, where $\alpha: C_n \to \mathbb{Z}$ is the augmentation and $a \in \bar{W}_{n-1}(C)$ and $b \in C_{n-1}$;
6. $D_n(a \otimes b) = 1_{n+1,W} \otimes (a \otimes b)$,
7. $D_i(a \otimes b) = D_i a \otimes D_i b, 0 \le i \le n-1$

Remarks. B.1.1. This differs from the original definition in that the last face-operator is "twisted" instead of the 0th. We have changed the definition so that we can write the path-space fibration as a twisted cartesian product using a twisted last face operator — this turns out to be the right thing to do when we want to write fibrations as (base)×(fiber) rather than (fiber)×(base) — see [10].

B.1.2. Note that $\bar{W}(C)_q = \mathbb{Z} \otimes C_0 \otimes \cdots \otimes C_{q-1}$. The ring structure on $\bar{W}(C)_q$ is defined so that multiplication is componentwise, on factors.

DEFINITION B.2. Let B and F be simplicial sets and let A be a simplicial group acting on F. Let $\xi: B_n \to F_{n-1}$ be a twisting function satisfying the identities: $\xi(F_n b) \cdot F_{n-1}\xi(b) = \xi(F_{n-1}b)$; $\xi(D_n b) =$ the unit of A_n if $b \in B_n$. We define $B \times_\xi F$, as follows:

1. as a simplicial set it is the cartesian product $B \times F$;
2. The face operators are given by $F_i(b, f) = (F_i b, F_i f)$, $i < n$, where $b \in B_n$, $f \in F_n$;
3. $F_n(b, f) = (F_n b, \xi(b) \cdot F_n f)$;
4. the degeneracy operators are defined as in the cartesian product.

Remarks. B.2.1. Twisted cartesian products are fibrations, in the context of simplicial sets — see [11].

B.2.2. The way we have defined the $\bar{W}$-construction implies that the function $a: \bar{W}_{q+1}(C) = \bar{W}_q(C) \otimes C_q \to C_q$ that simply sends $w \otimes c$ to c, is a twisting function. The corresponding twisted cartesian product turns out to be acyclic — it is the path space fibration of C. The contracting homotopy for (the chain-complex of) this space is ϕ, which sends $w \times c$, with $c \ne 1$ to $w \otimes c \times 1$, where w (and $w \otimes c) \in \bar{W}_q(C)$, and it sends elements of the form $w \times 1$ to 0.

B.2.3. Note that $\mathbb{Z}\pi$ can be regarded as an R-complex, concentrated in dimension 0. The main property of the $\bar{W}$-construction is that, as a *semi-simplicial complex*, $K(M, n) = \bar{W}(\mathbb{Z}\pi)$

PROPOSITION B.3. *Let* C *be an R-complex. Then the canonical m-structure,* $f_n: \mathcal{C}(\bar{W}(C) \times_a C) \to \mathrm{Hom}_{\mathbb{Z}S_n}(\mathrm{R}(S_n), (\mathcal{C}(\bar{W}(C) \times_a C))^n)$ *on* $\mathcal{C}(\bar{W}(C) \times_a C)$ *has the following property:* $f_n(w \times 1)(r) = (-1)^{\dim(w)}\mathfrak{H}_n \circ f_n(w \times 1)(\partial r) + \mathfrak{H}_n \circ f_n(a(w \times 1))(r)$, *where* $r \in \mathrm{R}(S_n)$, $w \in \bar{W}(C)$ *and* $\mathfrak{H}_n = 1 \otimes \cdots \otimes 1 \otimes \phi + 1 \otimes \cdots \otimes 1 \otimes \phi \otimes \epsilon + \cdots + \phi \otimes \epsilon \otimes \cdots \otimes \epsilon$, *where* $\epsilon: \bar{W}(C) \times_a C \to \mathbb{Z}$ *is the augmentation.*

PROOF. This follows from the way the canonical m-structure on the m-simplex was defined in the discussion following 4.1 of [24]. The chain-contraction, $\mathfrak{s}$, that was used there was a right inverse of the last-face operator, $\tilde{F}$, and this condition is also satisfied by the chain-contraction, ϕ, defined in remark 2 following B.2. □

Recall the result of V. K. A. M. Gugenheim in [10]:

Let $X \times_\xi F$ be a twisted cartesian product, where F is a left topological module over ΩX. Then there exists a twisting cochain $\bar{\xi}: X \to \Omega X$ with associated twisted tensor product, $X \otimes_{\bar\xi} F$ and a natural contraction $(\varphi_\xi, f_\xi, g_\xi): X \times_\xi F \to X \otimes_{\bar\xi} F$. The maps in this contraction (including the chain-homotopy) preserve the action of ΩX on F.

Let X be an R-complex. Then, as in [26], we use Gugenheim's result to construct a twisted tensor product $\bar W X \otimes_u X$ and a contraction $(\varphi_a, f_a, g_a): \bar W(X) \times_a X \to \bar W X \otimes_u X$. The contracting chain-homotopy of $\bar W(X) \times_a X$ can be transported to $\bar W X \otimes_u X$: we get a contracting homotopy $\hat\varphi = f_a \circ \phi \circ g_a: \bar W X \otimes_u X \to \bar W X \otimes_u X$. In [25] a simplified formula was given for in 2.4 (where it was called '$\mathfrak{s}$') that implies that it is self-annihilating. We can, consequently, use a procedure like that used in 3.1 to inductively construct a coherent m-structure on $\bar W X \otimes_u X$ — this will be denoted $\bar f_n: \mathcal{C}(\bar W(X) \times_a X) \to \mathrm{Hom}_{\mathbb{Z} S_n}(\mathrm{R}(S_n), (\mathcal{C}(\bar W(X \times_a X))^n)$. The inductive step involves defining $\bar f_n(b \otimes 1)(\sigma) = H_n \circ \bar f_n(\partial_\otimes (b \otimes 1))(\sigma) + (-1)^{\dim(b)} \bar f_n(b \otimes 1)(\partial \sigma)$, where $H_n = 1 \otimes \cdots \otimes 1 \otimes \hat\varphi + 1 \otimes \cdots \otimes 1 \otimes \hat\varphi \otimes \epsilon + \ldots$ The construction of $\{\bar f_n\}$ makes explicit use of the coherent m-structure on X (which exists since it is an R-complex, which means it is also a simplicial complex). This induces a coherent m-structure on X — i.e. the m-structure on $\bar W X \otimes_u X$ is compatible with the right-action of X, so we can form the quotient $\bar W X \otimes_u X \otimes_X \mathbb{Z} = X$. This induced m-structure generally will not agree with $\mathcal{C}(X)$. By abuse of notation, we will refer to this induced m-structure on X by $\bar f_n$.

PROPOSITION B.4. *Let X be an R-complex. Then there exists a complex equipped with:*

1. *a contraction $(u, v, \varphi): \widehat{W(X)} \to X$, where $v: \mathcal{C}(\bar W X) \to \widehat{W(X)}$ is a strict morphism of m-coalgebras;*

2. *a strict morphism of m-coalgebras $\iota: (\bar W X, \bar f_n) \to \widehat{W(X)}$.*

$\widehat{W(X)}$ *is functorial with respect to maps of R-complexes. Consequently the map* id: $\bar W X \to \bar W X$ *preserves these two m-structures up to a natural chain-homotopy.*

PROOF. We form the algebraic mapping cylinder, M, of $g: \bar W X \otimes_u X \to \bar W(X) \times_a X$. We construct a self-annihilating contracting chain-homotopy on M that restricts to on $\bar W X \otimes_u X$ and f on $\bar W(X) \times_a X$. Now we use this and lemma 4.1 on page 28 to build a coherent m-structure on M, using an argument like that used in 4.2 on page 30. By construction, this m-structure restricts to that of $(\bar W X, \bar f_n)$ on that summand. Proposition B.2 above implies that the coherent m-structure on M restricts to that of $\bar W(X) \times_a X$. on that summand. All of M comes equipped with a right X-action — it extends to the $\bar W(X) \times_a X$-summand because g_a is a homomorphism of right X-modules (this is proved in [10]). Consequently, we can form the quotient $M \otimes_X \mathbb{Z}$. This is the algebraic mapping cylinder of a map from $\bar W(X,) to \mathcal{C}(\bar W(X) \times_a X) \otimes_X \mathbb{Z}$. The original contraction $(\varphi_a, f_a, g_a): \bar W(X) \times_a X \to \bar W X \otimes_u X$ becomes a contraction $(\tilde\varphi_a, \tilde f_a, \tilde g_a): \mathcal{C}(\bar W(X) \times_a X) \otimes_X \mathbb{Z} \to \mathcal{C}(\bar W(X))$ — here we have used the fact that the natural chain-homotopy, φ_a, is also a homomorphism of right X-modules — see [7] for a proof.

Claim: The map $\tilde{f}_a$ is a strict morphism of m-structures.

This follows from the fact that it coincides with the projection $\bar{W}(X) \times_a X \rightarrow \bar{W}(X)$, which is a simplicial map.

The complex $\widehat{W(X)}$ is the result of taking the union of $M \otimes_X \mathbb{Z}$ with the algebraic mapping cylinder of along the common embedded copy of $(\bar{W}(X) \times_a X) \otimes_X \mathbb{Z}$. The algebraic mapping cylinder of trivially has a coherent m-structure since it is a homomorphism of coherent m-coalgebras (just take the m-structure of $(\bar{W}(X) \times_a X) \otimes_X \mathbb{Z} \otimes I$ union that of $\bar{W}(X)$). $\square$

THEOREM B.5. *Let* X *be an R-complex and let* $g: \bar{B}(X) \rightarrow \bar{W} X$ *be the natural map defined by Eilenberg and MacLane in [7]. Then there exists a complex equipped with:*

1. *a contraction* $(u, v, \varphi): \widehat{W(X)} \rightarrow \bar{W} X$, *where* $v: (\bar{W} X, \bar{f}_n) \rightarrow \widehat{W(X)}$ *is a strict morphism of m-coalgebras;*

2. *a strict morphism of m-coalgebras* $\iota: \bar{B}(X) \rightarrow \widehat{W(X)}$.

$\widehat{W(X)}$ *is functorial with respect to maps of R-complexes. Consequently* $g: \bar{B}(X) \rightarrow (\bar{W} X, \bar{f}_n)$ *preserves m-structures up to a natural chain-homotopy.*

PROOF. This is very similar to the proof of the preceding result. $\square$

THEOREM B.6. *Let* X *be an R-complex and let* $g: \bar{B}(X) \rightarrow \bar{W} X$ *be the natural map defined by Eilenberg and MacLane in [7]. Then there exists:*

1. *a contraction* $(\varphi, j, i): P(X) \rightarrow \bar{W} X$, *where* $i: \mathcal{C}(X) \rightarrow P(X)$ *is a homomorphism of m-coalgebras;*

2. *a homomorphism of m-coalgebras* $e: \bar{B}(X) \rightarrow P(X)$.

$P(X)$ *is functorial with respect to maps of R-complexes. Consequently* $g = \varphi \circ e: \bar{B}(X) \rightarrow \mathcal{C}(X)$ *preserves m-structures up to a natural chain-homotopy.*

We will combine this result with the preceding one to conclude that the map $g: \bar{B}(\Omega X) \rightarrow \mathcal{C}(X)$ preserves m-structures up to a natural chain-homotopy. In fact we can conclude the existence of a contraction (namely the composite of $\widehat{W(X)}$ and $\widehat{W(X)}$) $(\varphi, j, i): P(X) \rightarrow \mathcal{C}(X)$, where $i: \mathcal{C}(X) \rightarrow P(X)$ is a strict morphism of m-coalgebras, $P(X)$ is equipped with a coherent m-structure, and there exists an inclusion $e: \bar{B}(X) \rightarrow P(X)$ that is a strict morphism of m-coalgebras and such that $g = \varphi \circ e: \bar{B}(X) \rightarrow \mathcal{C}(X)$.

C. Proof of 2.12 (on page 43)

This essentially follows by the same argument that implies that the differential for the cobar-construction defined in 2.1 on page 38, is self-annihilating.

We will prove the result for $Z_{n,\infty}(\mathfrak{R})$. This will imply the result in general because the differential on $Z_{n,m}(\mathfrak{R})$ is induced by that in $Z_{n,\infty}(\mathfrak{R})$ (via $p_{\infty,m}$). It is only necessary to verify that

$$\hat{\mathfrak{T}}(\ker p_{\infty,m}) \subset \ker p_{\infty,m}$$

This follows immediately from the fact that $\ker p_{\infty,m}$ consists of elements of

$$\bigoplus \Sigma^{-k}\mathfrak{R}_k$$

of total rank $\geq m$, and each term of $\hat{\mathfrak{T}}$ increases rank.

Let $A(\alpha)_t = \sum_{j=1}^{t-1} \alpha_j$, where $A(\alpha)_0 = 0$ and define

(C.1) $$\hat{\mathfrak{T}}(\alpha)_{t,j} = \sum_{i=1+A(\alpha)_t}^{A(\alpha)_{t+1}} (-1)^{i+|\alpha|j+ij} \downarrow^{|\alpha|+j-1} \circ(\Delta_j \circ_i *) \circ \uparrow^{|\alpha|}$$

Then

$$\hat{\mathfrak{T}} = \sum_\alpha \sum_{j=1}^\infty \sum_{t=1}^n \hat{\mathfrak{T}}(\alpha)_t$$

PROPOSITION C.1. *If $\alpha(j)$ is the sequence that is identical to α, except that $j-1$ has been added to α_t, then it suffices to prove that $\sum_{j+j'=v} \hat{\mathfrak{T}}(\alpha(j))_{t,j'} \circ \hat{\mathfrak{T}}(\alpha)_{t,j} = 0$ for all t, α, and $v > 0$.*

PROOF. This is due to the fact that

$$\hat{\mathfrak{T}}(\alpha(j))_{t',j'} \circ \hat{\mathfrak{T}}(\alpha)_{t,j} = -\hat{\mathfrak{T}}(\alpha(j'))_{t,j} \circ \hat{\mathfrak{T}}(\alpha)_{t',j'}$$

for all t, t' with $t \neq t'$. This is due to the Commutativity Identity of a formal coalgebra (see 2.1 on page 11). The fact that $t \neq t'$ implies that the compositions with $\Delta_{j'}$ in the left factor are "out of the range" of the composition with Δ_j in the right factor, so the two compositions can be permuted via the Commutativity Identity. We verify this by direct computation:

$$\hat{\mathfrak{T}}(\alpha(j))_{t',j'} \circ \hat{\mathfrak{T}}(\alpha)_{t,j}$$

$$= \sum_{i'=1+A(\alpha(j))_{t'}}^{A(\alpha(j))_{t'+1}} (-1)^{i'+(|\alpha|+j-1)j'+i'j'} \downarrow^{|\alpha|+j+j'-2} \circ(\Delta_{j'} \circ_{i'} *) \circ \uparrow^{|\alpha|+j-1}$$

$$\circ \sum_{i=1+A(\alpha)_t}^{A(\alpha)_{t+1}} (-1)^{i+|\alpha|j+ij} \downarrow^{|\alpha|+j-1} \circ(\Delta_j \circ_i *) \circ \uparrow^{|\alpha|}$$

$$= \sum_{i'=1+A(\alpha(j))_{t'}}^{A(\alpha(j))_{t'+1}} \sum_{i=1+A(\alpha)_t}^{A(\alpha)_{t+1}}$$

$$(-1)^{i'+(|\alpha|+j-1)j'+i'j'+i+|\alpha|j+ij} \downarrow^{|\alpha|+j+j'-2}$$

$$\circ (\Delta_{j'} \circ_{i'} *) \circ (\Delta_j \circ_i *) \circ \uparrow^{|\alpha|}$$

Now if $t' < t$, then $i' < i$ and we compute the reverse composition

$$
= \sum_{i=1+A(\alpha(j'))_t}^{A(\alpha(j'))_{t+1}} \sum_{i'=1+A(\alpha)_{t'}}^{A(\alpha)_{t'+1}}
$$
$$
(-1)^{i+(|\alpha|+j'-1)j+ij+i'+|\alpha|j'+i'j'} \downarrow^{|\alpha|+j+j'-2}
$$
$$
\circ \left(\Delta_j \circ_{i''} * \right) \circ \left(\Delta_{j'} \circ_{i'} * \right) \circ \uparrow^{|\alpha|}
$$

which, by the Commutativity Identity, is equal to

$$
= \sum_{i''=1+A(\alpha(j'))_t}^{A(\alpha(j'))_{t+1}} \sum_{i'=1+A(\alpha)_{t'}}^{A(\alpha)_{t'+1}}
$$
$$
(-1)^{i''+(|\alpha|+j'-1)j+i''j+i'+|\alpha|j'+i'j'+jj'} \downarrow^{|\alpha|+j+j'-2}
$$
$$
\circ \left(\Delta_{j'} \circ_{i'} \left(\Delta_j \circ_{i''-(j'-1)} \Delta_{j'} \circ_{i'} * \right) \right) \circ \uparrow^{|\alpha|}
$$

Now we compare terms with the same type of composition-operation — we must have $i'' = i + j - 1$. The assumption that $t > t'$ implies that the indices that range of values of the subscript i'' will equal the range of the subscript i plus $j' - 1$. A straightforward (but tedious) computation shows that these terms have *opposite signs*.   □

Next we prove:

PROPOSITION C.2. *Under the hypotheses of this appendix,*

$$
\sum_{j+j'=v} \hat{\mathfrak{T}}(\alpha(j))_{t,j'} \circ \hat{\mathfrak{T}}(\alpha)_{t,j} = 0
$$

for all values of t such that $1 \leq t \leq n$ and all values of $v > 0$. We assume that j and j' are both ≥ 0.

The Symmetry Identity of a symmetric formal coalgebra (2.20 on page 18) implies that it suffices to prove that $\sum_{j+j'=v} \hat{\mathfrak{T}}(\alpha(j))_{1,j'} \circ \hat{\mathfrak{T}}(\alpha)_{1,j} = 0$ since the $\hat{\mathfrak{T}}(\alpha)_{t,j}$ are mapped into each other by suitable permutations:

$$
\hat{\mathfrak{T}}(\alpha)_{t,j} = G(j)^{-1} \circ \hat{\mathfrak{T}}(\alpha)_{1,j} \circ G
$$

where G is the permutation

$$
\begin{pmatrix}
1 & \cdots & A & A+1 & \cdots & A+\alpha_t & A+\alpha_t+1 & \cdots \\
\alpha_t+1 & \cdots & A+\alpha_t & 1 & \cdots & \alpha_t & A+\alpha_t+1 & \cdots
\end{pmatrix}
$$

and $G(j)$ is the permutation

$$
\begin{pmatrix}
1 & \cdots & A & A+1 & \cdots & A+\alpha_t+j-1 & A+\alpha_t+j & \cdots \\
\alpha_t+j & \cdots & A+\alpha_t+j-1 & 1 & \cdots & \alpha_t+j-1 & A+\alpha_t+j & \cdots
\end{pmatrix}
$$

We will prove this by direct computation:

$$\sum_{j+j'=m-1} \sum_{i'=1}^{\alpha'_1} (-1)^{i'+|\alpha'|j'+i'j'} \downarrow^{|\alpha|+j'-1} \circ (\Delta_{j'} \circ_{i'} *) \circ$$

$$\uparrow^{|\alpha'|} \sum_{i=1}^{\alpha_1} (-1)^{i+|\alpha|j+ij} \downarrow^{|\alpha|+j-1} \circ (\Delta_j \circ_i *) \circ \uparrow^{|\alpha|}$$

$$= \sum_{j+j'=m-1} \sum_{i'=1}^{\alpha_1+j-1} (-1)^{i'+(|\alpha|+j-1)j'+i'j'} \downarrow^{|\alpha|+j'+j} \circ (\Delta_{j'} \circ_{i'} *)$$

$$\circ \uparrow^{|\alpha|+j-1} \sum_{i=1}^{\alpha_1} (-1)^{i+|\alpha|j+ij} \downarrow^{|\alpha|+j-1} \circ (\Delta_j \circ_i *) \circ \uparrow^{|\alpha|}$$

$$= \sum_{j+j'=m-1} \sum_{i'=1}^{\alpha_1+j-1} \sum_{i=1}^{\alpha_1} (-1)^{i'+(|\alpha|+j-1)j'+i'j'+i+|\alpha|j+ij}$$

$$\downarrow^{|\alpha|+j'+j} \circ (\Delta_{j'} \circ_{i'} (\Delta_j \circ_i *)) \circ \uparrow^{|\alpha|}$$

Now we replace j' by k and collect terms with $j'+j=m-1$ — we can also replace j by $m-k+1$. We can eliminate extraneous factors of $\downarrow^{|\alpha|+j'+j} \circ$ and $\circ \uparrow^{|\alpha|}$.

$$= \sum_{j+j'=m-1} \sum_{i'=1}^{\alpha_1+j-1} \sum_{i=1}^{\alpha_1} (-1)^{i'+(|\alpha|+m-k)k+i'k+i+|\alpha|(m-k+1)+i(m-k+1)}$$

$$(\Delta_k \circ_{i'} (\Delta_{m-k+1} \circ_i *))$$

We can simplify the exponent of -1 somewhat:

$$(-1)^{|\alpha|} \sum_{k=1}^{m} \sum_{i'=1}^{\alpha_1+m-k} \sum_{i=1}^{\alpha_1} (-1)^{i'+(m-k)k+i'k+|\alpha|m+i(m-k)} (\Delta_k \circ_{i'} (\Delta_{m-k+1} \circ_i *))$$

Now we distinguish two cases: $i' < i$ and $i' \geq i$. In the first case we can apply identity 2b in 2.1 on page 11, to rewrite the equation in the form:

$$(-1)^{|\alpha|} \sum_{j+j'=m-1} \sum_{i'=1}^{\alpha_1+j-1} \sum_{i=1}^{\alpha_1} (-1)^{i'+(m-k)k+i'k+|\alpha|m+i(m-k)+k(m-k+1)}$$

$$(\Delta_{m-k+1} \circ_{i+k-1} \Delta_k \circ_{i'} *)$$

$$= (-1)^{|\alpha|} \sum_{j+j'=m-1} \sum_{i'=1}^{\alpha_1+j-1} \sum_{i=1}^{\alpha_1}$$

$$(-1)^{i'+i'k+|\alpha|m+i(m-k)+k} (\Delta_{m-k+1} \circ_{i+k-1} \Delta_k) \circ_{i'} *$$

The exponent of -1 is, in this case, $i'+i'k+|\alpha|m+i(m-k)+k$ (the term $k(m-k+1)$ comes from the Commutativity Identity in 2.1 on page 11). The corresponding term (i.e. with the same composition-operations and subscripts of the Δ's) from our original formula has a sign of -1 with the exponent equal to $i + k - 1 + (k - 1)(i' + m - k + 1) + (i + k - 1)(m - k + 1) + |\alpha|m$ and this simplifies (using mod 2 operations) to $i' + ki' + i(m - k) - k + 1 + |\alpha|m$. (In order to compute this, we perform a linear transformation on that sends k to $m - k + 1$, i' to $i + k - 1$, $m - k + 1$ to k, and i to i'). It is not hard to see that the parity of these two numbers will be different so the corresponding terms of the composition will cancel out.

In the second case we apply the Associativity Identity in 2.1 on page 11 to rewrite the equation in the form:

$$(-1)^{|\alpha|} \sum_{j+j'=m-1} \sum_{i'=1}^{\alpha_1+j-1} \sum_{i=1}^{\alpha_1} (-1)^{i'+(m-k)k+i'k+|\alpha|m+i(m-k)} \left(\Delta_k \circ_{i'} \left(\Delta_{m-k+1} \circ_i * \right) \right)$$

where we have used the identity 2a in the definition of a formal coalgebra. We can simplify the exponent of -1 somewhat to get:

$$(-1)^{|\alpha|} \sum_{j+j'=m-1} \sum_{i'=1}^{\alpha_1+j-1} \sum_{i=1}^{\alpha_1} (-1)^{i'+mk-k+(i+i')k+|\alpha|m+im} \left(\Delta_k \circ_{i'+i-1} \Delta_{m-k+1} \right) \circ_i *$$

Now we replace $i' + i$ by $m - k - l + 2$ (with $i' = m - k - \lambda + 2 - i$) to get

$$(-1)^{|\alpha|} \sum_{k=1}^{m} \sum_{\lambda=1}^{m-k} \sum_{i=1}^{\alpha_1} (-1)(-1)^{m-k-\lambda-i+mk-k+(m-k-\lambda)k+|\alpha|m+im}$$

$$\left(\Delta_k \circ_{m-k-\lambda+1} \Delta_{m-k+1} \right) \circ_i *$$

$$= (-1)^{|\alpha|} \sum_{k=1}^{m} \sum_{\lambda=1}^{m-k} \sum_{i=1}^{\alpha_1} (-1)^{m-\lambda-i+(-k-\lambda)k+|\alpha|m+im}$$

$$\left(\Delta_k \circ_{m-k-\lambda+1} \Delta_{m-k+1} \right) \circ_i *$$

— here we have performed some mod 2 simplifications on the exponent of -1. This further simplifies to:

$$= (-1)^{|\alpha|} \sum_{k=1}^{m} \sum_{\lambda=1}^{m-k} \sum_{i=1}^{\alpha_1} (-1)^{m-\lambda-i+k-\lambda k+|\alpha|m+im}$$

$$\left(\Delta_k \circ_{m-k-\lambda+1} \Delta_{m-k+1} \right) \circ_i *$$

Now we can break up the power of -1 and re-write the sums as:

$$= (-1)^{|\alpha|} \sum_{i=1}^{\alpha_1} (-1)^{m-i+|\alpha|m+im}$$

$$\sum_{k=1}^{m} \sum_{\lambda=1}^{m-k} (-1)^{k+\lambda+k\lambda} \left(\Delta_k \circ_{m-k-\lambda+1} \Delta_{m-k+1} \right) \circ_i *$$

The second line vanishes by the defining property of a formal $A(\infty)$-coalgebra.

D. The maps $\{\mathfrak{M}_n\}$

In this appendix we will prove that the maps $\{\mathfrak{M}_n\}$, defined in 3.9 (on page 65) and 3.10 (on page 65) have the properties mentioned in those results. Throughout this appendix we will make use of a $\mathbb{Z}$-bilinear operation, $\odot$, defined on $\mathrm{R}(S_n)$ as follows:

1. $a[a_1|\ldots|a_k] \odot b[b_1|\ldots|b_m] = a[a_1|\ldots|a_k \cdot b|b_1|\ldots|b_m]$;
2. $[] \odot [a_1|\ldots|a_k] = 0$;

We first consider the m-structure on $\mathcal{F}(\bar{\mathcal{B}}(C))$. We may assume that the formula in 3.6 on page 63 has the following form:

$$(\mathrm{D}.1) \quad \mathfrak{f}[\bar{\mathcal{B}}(C)]_n([c])([g_1|\ldots|g_k]) = (-1)^{(n-1)\cdot(\dim(c)+1)} g_1 \cdots g_{n-1} \circ H_n^{g_{n-1}^{-1}\cdots g_1^{-1}}$$

$$\circ H_n^{g_{n-1}^{-1}\cdots g_2^{-1}} \circ \cdots \circ H_n^{g_{n-1}^{-1}} \circ H_n \circ \mathfrak{f}[C]_n(c)([g_n|\ldots|g_k])$$

with no summation over k. This is due to the fact we are working with the m-structure of $\bar{\mathcal{B}}(C)$ rather than that of the twisted tensor product $\bar{\mathcal{B}}(C) \otimes_x C$ (the argument in 3.1 is designed to compute the m-structure of the twisted tensor product). All terms of the summation in 3.6 that result from a factor of 1 in $1 \otimes \otimes 1 \otimes S$ will give rise to factors of C that will either:

1. be equal to the 0-dimensional element 1, in which case they die due to the map $\downarrow$ as in the preceding remark or;
2. they will be of dimension > 0 in which case they die under the projection $\bar{\mathcal{B}}(C) \otimes_x C \to \bar{\mathcal{B}}(C)$.

It follows that, for our present purposes, $\mathfrak{f}[\bar{\mathcal{B}}(C)]_n(\bar{\mathcal{B}})([c])([g_1|\ldots|g_k])$ is nonzero if and only if the following n numbers are all distinct: $\{g_{n-1}^{-1}\cdots g_1^{-1}(n), g_{n-1}^{-1} \cdot g_2^{-1}(n), \ldots, g_{n-1}^{-1}(n), n\}$. The maps $\{\mathfrak{M}'_n\}$, defined in 3.9, have the following important property that we will use in the sequel:

$$(\mathrm{D}.2) \qquad \mathfrak{M}'_n([g_1|\ldots|g_n] \odot z) = \mathfrak{M}'_n([g_1|\ldots|g_n]) \odot z$$

PROPOSITION D.1. *The map $\mathfrak{M}'_n\colon K_n \to \mathrm{R}(S_n)$ is a chain-map for all $n \geq 1$.*

PROOF. We will prove this for canonical basis elements $[g_1|\ldots|g_k]$, $k \geq n-1$, $g_i \in S_n$. We first give an inductive definition of the boundary operation in $\mathrm{R}(S_n)$ as follows: $\partial[g_1|\ldots|g_t] \odot a = (\partial[g_1|\ldots|g_t]) \odot a - (-1)^t[g_1|\ldots|g_t] \odot \partial(a)$.

Case 1: $k = n$. Consider the set of $n+1$ numbers: $S = \{g_n^{-1}\cdots g_1^{-1}(n), g_n^{-1}\cdots g_2^{-1}(n), \ldots, g_n^{-1}(n), n\}$. We will use this in arguments regarding the quantities $\mathfrak{P}(g_1, \ldots, g_{n-1})$, define in 3.9, for elements of the boundary of $[g_1|\ldots|g_n]$:

1. $\mathfrak{P}(g_1, \ldots, g_{n-1})$ is nonzero if and only if the first n elements of the set S are distinct;
2. $\mathfrak{M}'_n$ of the ith term of $\partial[g_1|\ldots|g_n]$, $[g_1|\ldots|g_{i-1}|g_ig_{i+1}|g_{i+1}|\ldots|g_n]$, is nonzero if and only if the result of deleting the ith element of the set S consists of n distinct values;

CLAIM D.1.1. *If $\mathfrak{M}'_n([g_1|\ldots|g_n]) = 0$, then $\mathfrak{M}'_n(\partial[g_1|\ldots|g_n]) = 0$.*

Proof of claim: If $\mathfrak{M}'_n([g_1|\ldots|g_n])$ then there must be two elements of the first n elements of the set S that are the same, say the ith and the jth, with $i < j$. If these common values agree with a third value, then the statements above imply that all terms of $\mathfrak{M}'_n(\partial[g_1|\ldots|g_n])$ will be zero, and this implies the claim.

If these two terms represent the *only* duplication in the set S, then there will be precisely two nonzero term of $\mathfrak{M}'_n(\partial[g_1|\ldots|g_n])$ and we claim that they will cancel out. Then we compute the values of $\mathfrak{P}(g_1,\ldots,g_{i-1},g_i g_{i+1},g_{i+1},\ldots,g_n)$ and $\mathfrak{P}(g_1,\ldots,g_{j-1},g_j g_{j+1},g_{j+1},\ldots,g_n)$. They will differ only in the parity of the permutation of the set $\{1,\ldots,n-1\}$ mapping it into the ordered sets: $A = S-i^{\text{th}}$ term and $B = S-j^{\text{th}}$ term (where both elements are numerically equal). But the permutation carrying A into B consists of $j - i - 1$ transpositions, so that $\mathfrak{P}(g_1,\ldots,g_{i-1},g_i g_{i+1},g_{i+1},\ldots,g_n) = (-1)^{j-i-1} \cdot \mathfrak{P}(g_1,\ldots,g_{j-1},g_j g_{j+1},g_{j+1},\ldots,g_n)$. Since the corresponding terms of $\partial[g_1|\ldots|g_n]$ differ by a factor of $(-1)^{j-i}$, the results of plugging these two terms into $\mathfrak{M}'_n$ will have opposite signs. It is not hard to see that this means they will cancel out.

CLAIM D.1.2. *If* $\mathfrak{M}'_n([g_1|\ldots|g_n]) \neq 0$, *then* $\mathfrak{M}'_n(\partial[g_1|\ldots|g_n]) = \partial\mathfrak{M}'_n([g_1|\ldots|g_n])$.

Proof of claim: $\mathfrak{M}'_n([g_1|\ldots|g_n]) = (-1)^n g_1 \cdots g_{n-1}\mathfrak{P}(g_1,\ldots,g_{n-1})[g_n]$, and its boundary is $(-1)^n g_1 \cdots g_{n-1}\mathfrak{P}(g_1,\ldots,g_{n-1})(g_n[\,] - [\,])$.

The fact that $\mathfrak{M}'_n([g_1|\ldots|g_n]) \neq 0$ implies that the first n elements of the set S defined above are all distinct. This means that at least one of these n elements must equal n, the last element of S — suppose that this is the ith element of S. Then the claim above implies that there will be at most two nonzero terms of $\mathfrak{M}'_n(\partial[g_1|\ldots|g_n])$, namely the ith and the last. As was argued above, we can conclude that $\mathfrak{P}(g_1,\ldots,g_{i-1},g_i g_{i+1},g_{i+1},\ldots,g_n) = (-1)^{n-i-1}\mathfrak{P}(g_1,\ldots,g_{n-1},g_n)$. Since the corresponding terms of $\partial[g_1|\ldots|g_n]$ differ by $(-1)^{n-i}$, the two terms of $\mathfrak{M}'_n(\partial[g_1|\ldots|g_n])$ have opposite signs and it is not hard to verify that this will equal $(-1)^n g_1 \cdots g_{n-1}\mathfrak{P}(g_1,\ldots,g_{n-1})(g_n[\,] - [\,]) = \partial\mathfrak{M}'_n([g_1|\ldots|g_n])$.

Case 2: The general case. We make use of equation D.2. Suppose a is a canonical basis element of $R(S_n)$. Our two boundaries are:

1. $\partial(\mathfrak{M}'_n([g_1|\ldots|g_n] \odot a)) = (-1)^n g_1 \cdots g_{n-1}\partial(\mathfrak{P}(g_1,\ldots,g_{n-1})[g_n] \odot a)) = (-1)^n g_1 \cdots g_{n-1} \cdot g_n\mathfrak{P}(g_1,\ldots,g_{n-1})[\quad] \odot a - (-1)^n g_1 \cdots g_{n-1}\mathfrak{P}(g_1,\ldots,g_{n-1})[\quad] \odot a - (-1)^n g_1 \cdots g_{n-1}\mathfrak{P}(g_1,\ldots,g_{n-1})[g_n] \odot \partial(a)$. The first two terms are zero by property 2 of the $\odot$ operator, so we have $-(-1)^n g_1 \cdots g_{n-1}\mathfrak{P}(g_1,\ldots,g_{n-1})[g_n] \odot \partial(a)$.

2. $\mathfrak{M}'_n(\partial([g_1|\ldots|g_n]\odot a)) = \mathfrak{M}'_n(\partial[g_1|\ldots|g_n]\odot a - (-1)^n[g_1|\ldots|g_n]\odot\partial a)) = \mathfrak{M}'_n(\partial[g_1|\ldots|g_n] \odot a) - (-1)^n\mathfrak{M}'_n([g_1|\ldots|g_n] \odot \partial a)) = (-1)^n g_1 \cdots g_{n-1} \cdot g_n\mathfrak{P}(g_1,\ldots,g_{n-1})[\quad] \odot a - (-1)^n g_1 \cdots g_{n-1}\mathfrak{P}(g_1,\ldots,g_{n-1})[\quad] \odot a - (-1)^n g_1 \cdots g_{n-1}\mathfrak{P}(g_1,\ldots,g_{n-1})[g_n] \odot \partial(a)$, and this is the same result as before.

□

We will now show that the maps essentially preserve composition operations and so define a morphism of formal coalgebras. We begin with some inductive definitions:

DEFINITION D.2. The twisted shuffle product $\circledast$ can be defined inductively on canonical basis elements $[a_1|\ldots|a_m]$, $[b_1|\ldots|b_n]$ via $[a_1|\ldots|a_m] \circledast [b_1|\ldots|b_n] = ([a_1|\ldots|a_m] \circledast [b_1|\ldots|b_{n-1}]) \odot [b_n] + (-1)^n ([a_1|\ldots|a_{m-1}] \circledast [b_1|\ldots|b_n]) \odot [(b_1 \cdots b_n)^{-1} a_m (b_1 \cdots b_n)]$, where $a_i, b_j \in S_t$.

DEFINITION D.3. We can give the following recursive definition of the T-maps on canonical basis elements of $R(S_t)$: $\mathfrak{T}_{\alpha_1,\ldots,\alpha_t}([a_1|\ldots|a_m] \odot b) = \mathfrak{T}_{\alpha_1,\ldots,\alpha_t}([a_1|\ldots|a_m]) \odot [T_{(a_1 \cdots a_m)^{-1}\{\alpha_1,\ldots,\alpha_t\}}]$, where $a_1,\ldots,a_m, b \in S_t$.

PROPOSITION D.4. *The following identities hold:*

1. $\left(\underbrace{T_{1,\ldots,m,\ldots,1}}_{i^{\text{th}} \, position}(\sigma) \right)^{-1} (i+j) = \sigma^{-1}(i) + j$, *where* $\sigma \in S_n$, *and* $0 \leq j \leq m-1$;

2. $\left(\underbrace{T_{1,\ldots,m,\ldots,1}}_{i^{\text{th}} \, position}(\sigma) \right)^{-1} \mathcal{S}_{i-1}(\sigma') \circ \underbrace{T_{1,\ldots,m,\ldots,1}}_{i^{\text{th}} \, position}(\sigma) = \mathcal{S}_{\sigma^{-1}(i)-1}$, *where* $\sigma' \in S_m$, $\sigma \in S_n$.

3. $\left(\underbrace{T_{1,\ldots,m,\ldots,1}}_{i^{\text{th}} \, position}(\sigma) \right)^{-1} (n+m-1) =$

$$\begin{cases} \sigma^{-1}(n) & \text{if } i < n, \sigma^{-1}(n) < \sigma^{-1}(i) \\ \sigma^{-1}(n) + m - 1 & \text{otherwise} \end{cases}$$

(note that $i \leq n$), $\sigma \in S_n$.

4. parity $\left(\underbrace{T_{1,\ldots,m,\ldots,1}}_{i^{\text{th}} \, position}(\sigma) \right) = (-1)^{(\sigma^{-1}(i)+j)(m-1)} \, \text{parity}(\sigma)$.

PROOF. 1. In the first identity consider the definition of $\underbrace{T_{1,\ldots,m,\ldots,1}}_{i^{\text{th}} \, position}(\sigma)$ — form the ordered set $\{\sigma(1),\ldots,\sigma(n)\}$ and add $m+1$ to all elements numerically larger than i; insert numbers $i+1,\ldots,i+m-1$ following i (see 2.7 in [24]). Then $\left(\underbrace{T_{1,\ldots,m,\ldots,1}}_{i^{\text{th}} \, position}(\sigma) \right)^{-1}$ is the index of i in this sequence, and this is clearly the same as its index before we carried out these operations.

2. This follows by observing that $\left(\underbrace{T_{1,\ldots,m,\ldots,1}}_{i^{\text{th}} \, position}(\sigma) \right)^{-1}$ maps the ordered set $\{\sigma^{-1}(i),\ldots,\sigma^{-1}(i) + m - 1\}$ to the ordered set $\{i,\ldots,i+m-1\}$ without changing the relative order of the elements, and the fact that $\mathcal{S}_{i-1}(\sigma')$ preserves the complement of this set in the set $\{1,\ldots,n+m-1\}$.

3. This follows from the statement made in 1 above — if $i = n$ then $n + m - 1$ falls in the range of the m-1 elements inserted after i. If $i < n$ and $\sigma^{-1}(n) < i$ then it falls before these inserted elements. If $i \neq n$ and $\sigma^{-1}(n) \geq i$ then it also falls after these inserted elements.

4. To prove statement 4 we the consider the ordered set $\{\sigma(1), \ldots, \sigma(n)\}$. In a trivial permutation the ith element would have precisely $i - 1$ elements to its left — namely $\{1, \ldots, i-1\}$. In composing this with transpositions to make it the trivial permutation, we can move all of the $\sigma^{-1}(i) - 1$ elements to the left of the ith, to its right and then move back the $i - 1$ elements that belong there. This is a total of $(i + \sigma^{-1}(i) - 2)$ transpositions. After performing $T_{\underbrace{1, \ldots, m, \ldots, 1}_{i^{\text{th}} \text{ position}}}$

we will have to perform the same transpositions, except that these element will have moved $m - 1$ times (because the ith element of the ordered set will be expanded into a sequence of length m). Since all other elements of $T_{\underbrace{1, \ldots, m, \ldots, 1}_{i^{\text{th}} \text{ position}}}(\sigma)\{1, \ldots, n + m - 1\}$ will be in the same relative positions as

they had in $\{\sigma(1), \ldots, \sigma(n)\}$ the change in parity will be as stated.

□

We can clearly combine the two statements above with the definition of the composition-operation in $\mathfrak{S}$ given in 2.14 of [24] to get:

COROLLARY D.5. *We can give a recursive definition of the composition-operation on $\mathfrak{S}$. Let $[a_1| \ldots |a_u] \in R(S_m)$ and let $[b_1| \ldots |b_v] \in R(S_n)$ be canonical basis elements. Then:* $[a_1| \ldots |a_u] \circ_i [b_1| \ldots |b_v] = \{[a_1| \ldots |a_u] \circ_i [b_1| \ldots |b_{v-1}]\} \odot +(-1)^v \{[a_1| \ldots |a_{u-1}] \circ_i [b_1| \ldots |b_v]\} \odot [\ ]$.

THEOREM D.6. *The maps $\{\mathfrak{M}'_n\}$ satisfy the identity:* $\mathfrak{M}'_{n+m-i}([a_1| \ldots |a_{m-1}] \circ_i [b_1| \ldots |b_{n-1}]) = (-1)^{(m-1)(i-1)} \mathfrak{M}'_m([a_1| \ldots |a_{m-1}]) \circ_i \mathfrak{M}'_n([b_1| \ldots |b_{n-1}])$.

PROOF. Note that $\mathfrak{M}'_{n+m-1}([a_1| \ldots |a_{m-1}] \circ_i [b_1| \ldots |b_{n-1}])$ has a number of terms that are in a 1-1 correspondence with the terms of $[a_1| \ldots |a_{m-1}] \circ_i [b_1| \ldots |b_{n-1}]$ — which, in turn, are derived from the shuffle product in 2.14 of [24]:

$$\circ_i = \circledast(R(\mathfrak{S}_{i-1}) \otimes \underbrace{\mathfrak{T}_{1, \ldots, m, \ldots, 1}}_{i^{\text{th}} \text{ position}}): R(S_m) \otimes R(S_n) \to R(S_{m+n-1})$$

Let $i' = (b_1 \cdots _{n-1})i$, so that $i = (b_1 \cdots b_{n-1})^{-1} i'$, and $(b_1 \cdots b_j)^{-1} i = (b_{j+1} \cdots b_{n-1})i'$. Let $\mathfrak{T}_{\underbrace{1, \ldots, m, \ldots, 1}_{i^{\text{th}} \text{ position}}}([b_1| \cdots |b_{n-1}]) = [c_1| \ldots |c_{n-1}]$.

Then $c_j = T_{\underbrace{1, \ldots, m, \ldots, 1}_{(b_1 \cdots b_{j-1})^{-1} i}}(b_j) = T_{\underbrace{1, \ldots, m, \ldots, 1}_{(b_j \cdots b_{n-1})i'}}(b_j)$.   Furthermore

$c_j \cdots c_{n-1} = T_{\underbrace{1, \ldots, m, \ldots, 1}_{(b_j \cdots b_{n-1})i'}}(b_j \cdots b_{n-1})$, by 2.9 of [24].

Let $[w_1| \ldots |w_{m+n-2}]$ be a term of $[a_1| \ldots |a_{m-1}] \circ_i [b_1| \ldots |b_{n-1}]$. We will consider the evaluation of $\mathfrak{P}(w_1, \ldots, w_{m+n-2})$. We require that the elements $W =$

$\{w_{m+n-2}^{-1}\cdots w_1^{-1}(n+m-1),\ldots,w_{m+n-2}^{-1}(n+m-1), n+m-1\}$ all be distinct. All of the $\{w_j\}$ will either be of the form $T_{\underbrace{1,\ldots,m,\ldots,1}_{(b_j\cdots b_{n-1})i'}}(b_j)$ or of the form

$\mathcal{S}_{(b_{q+1}\cdots b_{n-1})i'-1}(a_{m-1})$. Furthermore, the product $w_{m+n-2}^{-1}\cdots w_1^{-1}$ will look like:

$$\left(T_{\underbrace{1,\ldots,i',\ldots,1}_{m^{\text{th}}\text{ position}}}(b_{n-1})\right)^{-1}\cdots\left(T_{\underbrace{1,\ldots,m,\ldots,1}_{(b_1\cdots b_{n-1})^{-1}i'}}(b_j)\right)^{-1}(\mathcal{S}_i(a_{m-1})^{-1}\cdots(\mathcal{S}_i(a_1)^{-1}$$

regardless of the order in which the shuffled terms occur, since a-terms are conjugated by b-terms that are shuffled to the left of them (in the twisted shuffle-product).

Claim: $\mathfrak{M}'_{n+m-1}([w_1|\ldots|w_{m+n-2}])$ is nonzero if and only if:

Case A $i = n$, $[w_1|\ldots|w_{m+n-2}] = [\mathcal{S}_i(a_1)|\ldots|\mathcal{S}_i(a_{m-1})|c_1|\ldots|c_{n-1}]$, and $\mathfrak{M}'_m([a_1|\ldots|a_{m-1}])$, $\mathfrak{M}'_n([b_1|\ldots|b_{n-1}])$ are both nonzero. Suppose k of the c-terms are shuffled to the left of the a-terms. Then the a-terms will be conjugated by these c-terms. By statement 2 of D.4, this results in $\mathcal{S}_i(a_1)$ being replaced by $\mathcal{S}_{(b_{k+1}\cdots b_{n-1})i'}(a_1)$. If $(b_{k+1}\cdots b_{n-1})i' \neq n$ then $\left(\mathcal{S}_{(b_{k+1}\cdots b_{n-1})i'}(a_1)\right)^{-1}(m+n-1) = m+n-1$ and the terms of $W = \{w_{m+n-2}^{-1}\cdots w_1^{-1}(n+m-1),\ldots, w_{m+n-2}^{-1}(n+m-1), n+m-1\}$ with as its rightmost factor $\left(\mathcal{S}_{(b_{k+1}\cdots b_{n-1})i'}(a_1)\right)^{-1}$ and the following term will be equal so that $\mathfrak{P}(w_1,\ldots,w_{m+n-2}) = 0$. On the other hand, if $(b_{k+1}\cdots b_{n-1})i' = n$ then there will be a duplication in values of the products of c-terms (after the a-terms have been stripped off) by statement 3 of D.4. Consequently, in this case $\mathfrak{P}(w_1,\ldots,w_{m+n-2}) = 0$. A similar argument implies that we cannot shuffle any c-term to the left of some (but not all) of the a-terms — there will exist terms of W that have these a-terms as their rightmost factor and the same argument used above will apply. It follows that the only term of the shuffle product that has a chance of giving a nonzero value of $\mathfrak{P}(w_1,\ldots,w_{m+n-2}) = 0$ is the one indicated. Statement 3 of proposition D.4 implies that in this case the set $W = \{w_{m+n-2}^{-1}\cdots w_1^{-1}(n+m-1),\ldots, w_{m+n-2}^{-1}(n+m-1), n+m-1\}$ will consist of
$\{B + (a_1\ldots a_{m-1})^{-1}(m) - 1, B + (a_2\ldots a_{m-1})^{-1}(m) - 1,\ldots, B + a_{m-1}^{-1}(m) - 1, B + m - 1, r_1(n),\ldots, r_{n-2}(n), n+m-1\}$,
where $B = (b_1\ldots b_{n-1})^{-1}(n)$ and

$$r_k(n) = \begin{cases} (b_{k+1}\cdots b_{n-1})^{-1}(n) & \text{if } (b_1\cdots b_k)^{-1}(n) < n, \\ & \text{and } (b_{k+1}\cdots b_{n-1})^{-1}(n) < B \\ (b_{k+1}\cdots b_{n-1})^{-1}(n) + m - 1 & \text{otherwise} \end{cases}$$

It follows that the set W will consist of distinct elements only if the sets $\{a_{m-1}^{-1}\cdots a_1^{-1}(m),\ldots, a_{m-1}^{-1}(m), m\}$ and $\{b_{n-1}^{-1}\cdots b_1^{-1}(n),\ldots, b_{n-1}^{-1}(n), n\}$ each consist of distinct elements. Suppose this is true. We will now compute $\wp(w_1,\ldots,w_{n+m-2})$ where $\wp(*)$ is defined in 3.9, statement 1b.

Note that the relative order of the $\{r_1(n), \ldots, r_{n-2}(n), n + m - 1\}$ is the same as that of $\{b_{n-1}^{-1} \cdots b_1^{-1}(n) = B, \ldots, b_1^{-1}(n), n\}$, except that the element B has been expanded to a sequence of m consecutive numbers, and the permutation has essentially been multiplied by the permutation $S_{B-1} \wp(a_1, \ldots, a_{m-1})$. We claim that $\wp(w_1, \ldots, w_{n+m-2}) = S_{B-1} \wp(a_1, \ldots, a_{m-1}) \cdot \underbrace{T_{1,\ldots,m,\ldots,1}}_{B^{\text{th}} \text{ position}}(\wp(b_1, \ldots, b_{n-1}))$. This means that the parities satisfy the condition:

$$\text{parity}(\wp(w_1, \ldots, w_{n+m-1}))$$
$$= (-1)^{(m-1)(B+1)} \text{parity}(\wp(b_1, \ldots, b_{n-1})) \cdot \text{parity}(\wp(a_1, \ldots, a_{m-1}))$$

— see statement 4 of 3.9.   The parity of $\underbrace{T_{1,\ldots,m,\ldots,1}}_{i^{\text{th}} \text{ position}}(b_1 \cdots b_{n-1})$

is $(-1)^{(B+i)(m-1)}$, by statement 4 of 3.9.   This implies that that $\mathfrak{M}'_{n+m-1}([a_1|\ldots|a_{m-1}] \quad \circ_i \quad [b_1|\ldots|b_{n-1}]) = (-1)^{(m-1)(i-1)} \mathfrak{M}'_m([a_1|\ldots|a_{m-1}]) \circ_i \mathfrak{M}'_n([b_1|\ldots|b_{n-1}])$

Case B $i < n$, there exists a value of q such that $(b_{q+1} \cdots b_{n-1})i' = n$,

$$[w_1|\ldots|w_{m+n-2}] = [c_1|\ldots|c_q|S_{(b_{q+1}\cdots b_{n-1}i')}(a_1)$$
$$|\ldots|S_{(b_{q+1}\cdots b_{n-1}i')}(a_{m-1})|c_{q+1}|\ldots|c_{n-1}]$$

and $\mathfrak{M}'_m([a_1|\ldots|a_{m-1}])$, $\mathfrak{M}'_n([b_1|\ldots|b_{n-1}])$ are both nonzero. Recall that statement 2 of D.4 implies that $(c_{q+1} \cdots c_{n-1})^{-1} S_{i'}(a_j) c_{q+1} \cdots c_{n-1} = S_{(b_{q+1}\cdots b_{n-1})i'}(a_j)$. In general $(b_{q+1} \ldots b_{n-1})i'$ is always $\leq n$ and if it is strictly $< n$ then $S_{(b_{q+1}\cdots b_{n-1})i'-1}(a_{m-1})$ leaves $n + m - 1$ fixed — and we can conclude that $\mathfrak{P}(w_1, \ldots, w_{m+n-2}) = 0$ by an argument like that used in case A above. A similar argument implies that no additional c-terms can be shuffled to the left of any of the a-terms. On the other hand, if there exists no value of q fitting the condition above it is not hard to see that $\mathfrak{P}(w_1, \ldots, w_{m+n-2}) = 0$.
$$W = \{w_{m+n-2}^{-1} \cdots w_1^{-1}(n+m-1), \ldots, w_{m+n-2}^{-1}(n+m-1), n+m-1\}$$

$$\{B + m - 1, r_1(n), \ldots, r_{q-1}(n), B + (a_1 \ldots a_{m-1})^{-1}(m) - 1,$$
$$B + (a_2 \ldots a_{m-1})^{-1}(m) - 1, \ldots, B + a_{m-1}^{-1}(m) - 1,$$
$$r_q(n), \ldots, r_{n-2}(n), n + m - 1\}$$

where $B = (b_1 \ldots b_{n-1})^{-1}(n)$ and

$$r_k(n) = \begin{cases} (b_{k+1} \cdots b_{n-1})^{-1}(n) & \text{if } (b_1 \cdots b_k)^{-1}(n) < n, \\ & \text{and } (b_{k+1} \cdots b_{n-1})^{-1}(n) < B \\ (b_{k+1} \cdots b_{n-1})^{-1}(n) + m - 1 & \text{otherwise} \end{cases}$$

so that $\wp(w_1, \ldots, w_{n+m-2}) = S_{B-1} \wp(a_1, \ldots, a_{m-1}) \cdot \underbrace{T_{1,\ldots,m,\ldots,1}}_{B^{\text{th}} \text{ position}}(\wp(b_1, \ldots, b_{n-1}))Z^{-1}$, where Z is the cyclic permutation

$$\begin{pmatrix} 1 & \cdots & m-1 & m & \cdots & q+m-1 \\ q+1 & \cdots & q+m-1 & 1 & \cdots & q \end{pmatrix}.$$ The parity

of $\wp(w_1, \ldots, w_{n+m-2})$ is equal to $(-1)^{(m-1)(B+1)} \cdot \text{parity}(Z)$. The parity of Z is equal to $(-1)^{q(m-1)}$, so that the parity of $\wp(w_1, \ldots, w_{n+m-2}) = (-1)^{(m-1)(B+q+1)}$. The parity of $T_{\underbrace{1, \ldots, m, \ldots, 1}_{i^{\text{th}} \text{ position}}}(b_1 \cdots b_{n-1})$ is $(-1)^{(B+i)(m-1)}$ and the product is equal to

$(-1)^{(m-1)(i+q-1)}$. The fact that we have selected the $q+1^{\text{st}}$ term of the shuffle product (i.e., we have shuffled all of the a-terms past q of the b-terms) we get an additional sign of $(-1)^{q(m-1)}$. Multiplying these factors together gives the same result as in case A. $\square$

Note that the formula D.1 can be written as:

$\mathcal{F}(f_n(\bar{B}), n)([c])([g_1|\ldots|g_k]) = (-1)^{(n-1)(\dim(c))} S \otimes \cdots \otimes S \circ f_n(c)(\mathfrak{M}'_n([g_1|\ldots|g_k])) = S \otimes \cdots \otimes S \circ \text{Hom}_{\mathbb{Z} S_n}(\mathfrak{M}'_n, 1) \circ f_n(c)([g_1|\ldots|g_k])$ (by the Koszul convention, since we have essentially permuted c and the map $\mathfrak{M}'_n$, which is of degree $-(n-1)$).

COROLLARY D.7. *The maps $\{\mathfrak{M}_n\}$ satisfy the identity:*

$\mathfrak{M}'_{n+m-1}([a_1|\ldots|a_{m-1}] \circ_i [b_1|\ldots|b_{n-1}])$

$$= (-1)^{(m-1)(i-1)} \mathfrak{M}_m([a_1|\ldots|a_{m-1}]) \circ_i \mathfrak{M}_n([b_1|\ldots|b_{n-1}])$$

PROOF. This is proved by induction, using D.5 and D.6. $\square$

E. The composition-operations of $Y_{n,m}(\mathfrak{R}) \bigstar_{\{\ell, \hat{\mu}_i\}} \mathcal{F}(\mathfrak{R}, n)$

Throughout this section we will assume that $(C, \{f[C]_n : C \rightarrow \text{Hom}_{\mathbb{Z} S_n}(\mathfrak{R}[C]_n, C^n)\})$ is a weakly-coherent m-coalgebra. We begin with:

DEFINITION E.1. Let $\bar{\alpha} = (\bar{\alpha}_1, \ldots, \bar{\alpha}_m)$ be a sequence of integers ≥ 0. Then $\mathcal{F}(\mathfrak{R}[C], \bar{\alpha}, m) \subset \mathcal{F}(\mathfrak{R}[C], m)$ is defined to be the sub-module of elements whose image is in $(\Sigma^{-1}C)^{\bar{\alpha}_1}) \otimes \cdots \otimes (\Sigma^{-1}C)^{\bar{\alpha}_m}) \subset \mathcal{F}(C) \otimes \cdots \otimes \mathcal{F}(C)$ (m copies).

DEFINITION E.2. Let $\mathfrak{R}$ be an f-resolution. Let $z \in Y(\mathfrak{R}, \alpha, n)$ and let $f = [r_1|\ldots|r_{\alpha_i}] \in \mathcal{F}(\mathfrak{R}, \bar{\alpha}, m)$, where $\bar{\alpha}$ is a sequence of integers ≥ 0. Then it is possible to define the composite

$$f \circ_{n+\sum_{j=1}^i \alpha_j} z$$

as

$$r_1 \circ_t r_2 \circ_{t+1} \cdots \circ_{t+\alpha_i-1} z$$

where $t = n + \sum_{j=1}^i \alpha_j$.

Explanation. If $\mathfrak{R} = \mathfrak{R}[C]$, the elements of $Y(\mathfrak{R}[C], \alpha, n)$ represent maps with targets in $C^n \otimes \bigotimes_j (\Sigma^{-1}C)^{\alpha_i}$, and the elements, $[r_1|\ldots|r_{\alpha_i}]$, of $\mathcal{F}(\mathfrak{R}[C], \bar{\alpha}, m)$ represent maps of $(\Sigma^{-1}C)^{\alpha_i} \subset \mathcal{F}(C)$, where each r_j corresponds to a map of a separate copy of $\Sigma^{-1}C$.

THEOREM E.3. *The coordinate coalgebra of* $C \otimes_\ell \mathcal{F}(C)$ *is* $Y_{n,m}(\mathfrak{R}[C]) \otimes \mathcal{F}(\mathfrak{R}[C], n)$
with a twisted composition-operation defined as follows:

Let $y \in Y(\mathfrak{R}[C], \alpha, n), f \in \mathcal{F}(\mathfrak{R}[C], n), y' \in Y(\mathfrak{R}[C], \beta, m), f' \in \mathcal{F}(\mathfrak{R}[C], \bar{\alpha}, m).$
Then $(y' \otimes f') \circ_j (y \otimes f) = \hat{y} \otimes (f' \circ_j f)$, *where* $\hat{y} \in Y(\mathfrak{R}[C], \gamma, n + m)$, *with*

$$\gamma_i = \begin{cases} \alpha_i & \text{if } i < j \\ \beta_i + \bar{\alpha}_i & \text{if } j \leq i < j + m \text{ and } \hat{y} \text{ is equal to the composite:} \\ \alpha_{i-j+1} & \text{if } i \geq j + m \end{cases}$$

$$\downarrow^{|\gamma|-1} \circ \mathcal{Z}\{\{\beta_1, \ldots, \beta_m\}, (\bar{\alpha}_1, \ldots, \bar{\alpha}_m)\} \cdot \mathcal{Z}\{\{n+m, |\beta|\}, (\sum_{i=1}^{j-1} \alpha_i, \sum_{i=j}^{n} \alpha_i)\}$$

$$\cdot (f' \circ_{n+m+|\beta|\sum_{i=1}^{j} \alpha_i} |\mathcal{Z}\{(j-1+m, n-j), (|\beta|, |\alpha|)\}|^{-1}$$

$$\cdot (\uparrow^{|\beta|-1})(y') \circ_j (\uparrow^{|\alpha|-1}(y))$$

REMARK. E.3.1. We will only indicate the "idea" of this formula:

1. The term $\uparrow^{|\beta|-1}(y') \circ_j \uparrow^{|\alpha|-1}(y)$ represents the portion of this composition
   that involves plugging a map $C \to C^m \otimes \mathcal{F}(C)^m$ into the image of a copy of
   C in the target of y. This composition results in a map $C \to C^{j-1} \otimes C^m \otimes$
   $\mathcal{F}(C)^m \otimes C^{n-j} \otimes \mathcal{F}(C)^n$ and;

2. the term $|\mathcal{Z}\{(j-1+m, n-j), (|\beta|, |\alpha|)\}|^{-1}$ shuffles this target to $C^{j-1} \otimes C^m \otimes$
   $C^{n-j} \otimes \mathcal{F}(C)^m \otimes \mathcal{F}(C)^n$ which really has $|\beta|$ copies of $\Sigma^{-1}C$ and $\mathcal{F}(C)^n$
   has $|\alpha|$ copies of $\Sigma^{-1}C$. The "absolute value" is taken because de-suspended
   copies of C are being shuffled past *ordinary* (i.e. non de-suspended) copies of
   C so there is no change in sign;

3. the term $\mathcal{Z}\{(n+m, |\beta|), (\sum_{i=1}^{j-1} \alpha_i, \sum_{i=j}^{n} \alpha_i)\}$ shuffles the first $j - 1$ copies
   of $\mathcal{F}(C)$ from the original target of y past the result of y', i.e. we get $C^{j-1} \otimes$
   $C^m \otimes C^{n-j} \otimes \mathcal{F}(C)^{j-1} \otimes \mathcal{F}(C)^m \otimes \mathcal{F}(C)^{n-j+1}$. The factor of $\mathcal{F}(C)^m$ was
   produced from y' — it represents a "perturbation" of the map f'.

4. the term $f' \circ_{n+m+|\beta|\sum_{i=1}^{j} \alpha_i} *$ is the result of composing the $\mathcal{F}(C)^n$-factor
   of the target of y (which was $C^n \otimes \mathcal{F}(C)^n$), with f'. The subscript $n + m +$
   $|\beta| \sum_{i=1}^{j} \alpha_i$ is used because:
   a. we need $n + m$ to move past the copies of $C^m \otimes C^n$ on the left end of
      $\uparrow^{|\beta|-1}(y') \circ_j \uparrow^{|\alpha|-1}(y)$;
   b. we need $|\beta|$ to move past the copies of $\mathcal{F}(C)^m$;
   c. we need $\sum_{i=1}^{j} \alpha_i$ to form the composition with the proper copies of
      $\Sigma^{-1}C$ in $\mathcal{F}(C)^n$.

5. the term $\mathcal{Z}\{(\beta_1, \ldots, \beta_m), (\bar{\alpha}_1, \ldots, \bar{\alpha}_m)\}$ shuffles the "perturbation",
   $\mathcal{F}(C)^m$, with the image of f' (the ith copy of $\mathcal{F}(C)$). These two sequences of
   copies of $\mathcal{F}(C)$ are supposed to be multiplied together using the product
   operation of $\mathcal{F}(C)$. Since multiplication in $\mathcal{F}(C)$ is nothing but taking
   the tensor product, this multiplication is accomplished by the shuffling
   operation above, and the definition $\gamma_i = \beta_i + \bar{\alpha}_i$.

6. A prolonged diagram chase shows that this definition of composite is compatible with the adjoint to the structure map defined in 3.2 on page 94:

$$\circ_i \colon Y_{n,m}(\mathfrak{R}[C]) \otimes \mathcal{F}(\mathfrak{R}[C], n) \rightarrow \operatorname{Hom}_{\mathbb{Z}}(C \otimes_{\ell} \mathcal{F}(C), (C \otimes_{\ell} \mathcal{F}(C))^n).$$

F. Calculations

We will recursively apply the algorithm described in the remarks following 2.27 on page 55. Since the homotopy kills off elements of $\mathrm{R}(S_n)$ with of the form $1 \cdot [\cdots]$, many terms of the formula in 2.27 contribute 0 (for instance all of the terms with the $\mathfrak{T}$-map). We will compute a homomorphism of DGA-coalgebras $h_n \colon \mathrm{R}(S_n) \rightarrow \mathcal{F}(\mathrm{R}(S_n), n)$ that is a right-inverse to the map $\mathfrak{M}_n \colon \mathcal{F}(\mathrm{R}(S_n), n) \rightarrow \mathrm{R}(S_n)$ defined in 3.10 on page 65. The algorithm in 2.27 (on page 55) does not uniquely determine such a map — it only gives an iterative procedure for constructing such maps. In each step of the iteration one may add an arbitrary boundary to the map constructed up to that point. We will use this idea to construct a map with the desired properties.

PROPOSITION F.1. *The only z-functions that play a significant role in the map $h_n \colon \mathrm{R}(S_n) \rightarrow \mathcal{F}(\mathrm{R}(S_n), n)$ in $\mathfrak{M}_n \circ h_n$ are of the form $z_{i_1,\ldots,i_n}$, where the i_j are all 0 or 1 (for the time being we will restrict our computations of the z_α to terms of the form $z_{1,1}, z_{1,1,1}, \ldots$). The map $\mathfrak{M}_n \circ$ **other terms** will vanish.*

PROOF.   1. In the formula for H_n (in 3.6 on page 63), we can ignore all terms but the first — i.e. H_n is effectively equal to $1 \otimes \cdots \otimes 1 \otimes S$ — see 3.6 and the discussion preceding it for the notation. This is due to the fact that all factors of the f_n-functions of $\bar{\mathcal{B}}(C)$ will be plugged into $\downarrow^{|\alpha|}$ for some value of $\alpha > 0$, and $\downarrow$ annihilates any element of dimension 0 — see 2.8 on page 42 and the remarks following it[1].

Recall, from 3.9 on page 65, that $\mathfrak{M}_n([g_1 | \ldots | g_k])$ is nonzero if and only if the quantities $\{g_{n-1}^{-1} \cdots g_1^{-1}(n), \ldots, g_{n-1}^{-1}(n), n\}$ are all distinct. We will show that this requirement is violated for the value of z_α whenever the sequence, α, has a term greater than 1.

The conclusion follows from considering the formula for the z_α in 2.27 on page 55. There are two significant terms in this formula:

a. $\mathcal{Z}\{\beta_1, \ldots, \beta_i\} \circ t_i \circ z_{\beta_1} \otimes \cdots \otimes z_{\beta_i} \circ \Delta_{\mathfrak{M}}^{i-1}$ and

b. $\sum_{i=1}^{|\alpha|} \sum_{j=2}^{\infty} (-1)^{i+|\alpha|j+ij} \Delta_j \circ_i z_{\alpha(i,j)}$

In the first case, we essentially have a kind of shuffle product of z_α's in which the z_{β_i}-term has had its indices translated up by $\sum_{j=1}^{i-1} |\beta_j|$ (so that the permutations contained in the factors of the shuffle product operate over disjoint ranges). In a given term of this shuffle product suppose a factor of z_{β_k} with $k < i$ gets shuffled to the right end of the entire term. Then the conclusion follows for this term since the permutation in stabilizes the number n so that the numbers $\{g_{n-1}^{-1} \cdots g_1^{-1}(n), \ldots, g_{n-1}^{-1}(n), n\}$ will not all be distinct — $g_{n-1}^{-1}(n) = n$. The general case follows by a similar argument: if the last t factors in a term of z_α are all taken from z_{β_i} (so they may move n around) the $n - t^{\text{th}}$ factor will stabilize all of the indices in the range of z_{β_i} so that $g_{n-1}^{-1} \cdots g_{n-t+1}^{-1}(n) = g_{n-1}^{-1} \cdots g_{n-t}^{-1}(n)$ and the conclusion now follows for all terms generated from part a of the formula in 2.27 on page 55.

---

[1] This is not to say factors of 1 never occur in $\mathcal{F}(\bar{\mathcal{B}}(C))$ — they only occur as a result of having a subscript of 0. Even in this case all factors of $\mathcal{F}(\bar{\mathcal{B}}(C))$ get plugged into $\downarrow$.

A similar conclusion follows for part b by induction — it is not hard to see that if the claim holds for $z_{\alpha(i,j)}$, then it will also hold for $\Delta_j \circ_i z_{\alpha(i,j)}$ — at least in the case we are dealing with (where $\Delta_j = 0$ for $j > 2$ and $\Delta_2 \circ_i z_{\alpha(i,j)} = \mathfrak{T}_{\underbrace{1,\ldots,2,\ldots,1}_{i^{\text{th}} \text{ position}}}(z_{\alpha(i,j)})$ — see [24]). This completes the proof

of the claim.

$\square$

The h_2 is defined on $[\,]$ by setting $z_{1,1}([\,]) = (1,2)[(1,2)]$, $z_{1,1,1}([\,]) = z_{1,1}([\,]) \otimes 1 \circledast \mathfrak{T}_{2,1}(z_{1,1}([\,]))$, etc. That this is well-defined follows from 2.13 on page 45 which implies that any procedure for computing $z_{1,\ldots,1}$ will, at least, be homotopic to what is stated above[2] — and one may vary the result of any stage of any computation in 2.12 by a boundary. The calculations in 3.6 imply that the definition of $z_{1,1}([\,])$ has the required property: we get $\mathfrak{M}_2([(1,2)]) = (1,2)$.

The construction of h_n for values of $n > 2$ is based upon the following lemma — which essentially implies that we can prescribe the values of the composite $f(x)^n \circ \mathcal{F}(f_n(\bar{B}), n)$:

LEMMA F.2. *Let $a = [g_1|\ldots|g_k] \in R(S_k)$ be an arbitrary basis element of dimension ≥ 1. Let $v_n(a) = (-1)^n[(1,n)|(1,n,2)|(2,n,3)|\ldots|(n-2,n,n-1)|(n-1,n)|a$. Then*

1. $\mathfrak{M}_n \circ \partial \circ v_n(a) = a$;
2. *If $e = [g_1|\ldots|g_{n-1}]$, and $\mathfrak{M}_n(e)$ is nonzero and $b = g_1 \cdots g_{n-1}$, then $\mathfrak{M}_n(\partial \circ v) = \mathfrak{M}_n(e)$, where $v = b[b^{-1}|e$.*

PROOF. This follows by direct computation, using remark 3 above. Essentially, $(n-1,n)$ maps n to $n-1$, $(i,n,i+1)\cdots(n-1,n) = (i,n)$ mapping n to i. When we take the boundary of $v_n(a)$, we get $(1,n)[(1,n,2)|(2,n,3)|\ldots|(n-1,n)|a - [(1,n) \cdot (1,n,2)|\ldots|(n-1,n)|a + \cdots$.

The first term results in a (when $\mathfrak{M}_n$ is applied) and all of the succeeding terms result in 0 since they contain the product $g_{n-1}^{-1} \cdots g_1^{-1}$, which is equal to the identity in this case. The second statement is clear: the boundary of v will have a term exactly equal to e — the first term. All other terms except the last will have the product of all of the elements equal to the identity element, and this will prevent them from contributing to the result since the first element of the set $\{g_{n-1}^{-1} \cdots g_1^{-1}(n), \ldots, g_{n-1}^{-1}(n), n\}$ will be equal to the last (n). The last term of the boundary will have a corresponding set of numbers that is missing g_{n-1}^{-1} — the first $n-2$ entries will be the product of the original set by g_{n-1}. We claim that this term contributes nothing as well since g_{n-1} must map some element of the set $\{1, \ldots, n-1\}$ to n itself — otherwise it would map n to itself (by the pigeonhole principal) and this would contradict the assumption that $\mathfrak{M}_n(e)$ is nonzero. $\square$

The main result of this appendix now follows easily — simply compute $z_{1,\ldots,1}(a)$ for any $a \in R(S_k)$ and plug it into $\mathfrak{M}_n$ — if the result isn't equal to a compute $v = v_n(\textbf{the difference})$ and vary $z_{1,\ldots,1}(a)$ by ∂v to get the desired result.

We conclude this appendix with some sample computation. The simple case corresponds to a straightforward application of formula 2.27 on page 55. The canonical

---

[2]Since it proved that the $Z_{n,m}(\mathfrak{R})$ are acyclic.

case corresponds to the case dealt with in this appendix — the construction of a right inverse to the $\mathfrak{M}_n$.

Simple $z_\alpha = \epsilon$, if α is a string of 0's with a single 1 in it.

1. $z_{1,1}([\,]) = -[(1,2)]$; in fact $z_\alpha([\,]) = -[(1,2)]$, where α is any string of 0's with two 1's in it.
2. $z_{1,2}([\,]) = -[(1,2)|(2,3)]$;
3. $z_{2,1}([\,]) = [(2,3)|(1,2)]$;
4. $z_{2,2}([\,]) = [(1,3,2)|(3,4)|(2,3)] + [(2,3)|(1,2)|(3,4)] - [(2,3)|(3,4)|(1,2)] - [(2,3,4)|(1,2)|(2,3)]$;

Canonical   1. $z_{1,1}([\,]) = -(1,2)[(1,2)]$;

2. $z_{1,2}([\,]) = -[(1,2,3)|(2,3)] - [(1,2)|(1,2)] + [(1,2,3)|(1,3,2)]$;
3. $z_{2,1}([\,]) = +[(2,3)|(2,3)] + [(1,3,2)|(1,2)] - [(1,3,2)|(1,2,3)]$
4. $z_{2,2}([\,]) = +[(1,3,2)|(3,4)|(3,4)] - [(1,3,2)|(2,4,3)|(2,3,4)] + [(1,3,2)|(2,4,3)|(2,3)] + [(1,3,4,2)|(1,2)|(3,4)] - [(1,3,4,2)|(3,4)|(1,2)] - [(2,3,4)|(1,2)|(1,2)] + [(2,3,4)|(1,2,3)|(1,3,2)] - [(2,3,4)|(1,2,3)|(2,3)]$;
5. $z_{1,3}([\,]) = +[(1,2)|(2,3,4)|(3,4)] - [(1,2)|(2,3,4)|(2,4,3)] + [(1,2)|(2,3)|(2,3)]$;
6. $z_{3,1}([\,]) = -[(3,4)|(2,3)|(2,3)] + [(3,4)|(1,3,2)|(1,2,3)] - [(3,4)|(1,3,2)|(1,2)]$;

These calculations were performed by a Pascal program. The author will provide the source code upon request and:

- It is available via anonymous ftp from `mcs.drexel.edu` (in the directory `pub/top`).
- It is available via `gopher` from `mcs.drexel.edu` (in the `top` directory)

Here is a listing of this program:

```
program tmap (input, output);
  const
    psize = 50;
  type
    permat = array[1..psize] of integer;
    perm = record
        size: integer;
        data: permat
      end;
    numseq = record
        size: integer;
        data: permat
      end;
    coeftype = record
        numval: integer;
        grpelt: perm
      end;
    barptr = ^barelt;
    barelt = record
        first: perm;
        rest: barptr
      end;
    bardatptr = ^bardata;
```

```pascal
bardata = record
    coef: coeftype;
    term: barptr;
    rest: bardatptr
  end;

var
  t1: bardatptr;
  i, j: integer;
  resp: char;
  mode: (simple, canonical);
procedure putpos (ind: integer);
begin
  if ind < 10 then
    write(chr(ind + ord('0')))
  else
    begin
      putpos(ind div 10);
      write(chr(ind mod 10 + ord('0')));
    end
end; {putpos}
function vsign (i, j: integer): integer;
begin
  if odd(i * j) then
    vsign := −1
  else
    vsign := 1
end;
procedure putnum (ind: integer);
begin
  if ind < 0 then
    begin
      write('−');
      putpos(abs(ind));
    end
  else
    putpos(ind)
end; {putnum}
function max (i, j: integer): integer;
begin
  if i >= j then
    max := i
  else
    max := j
end; {max}
procedure vperm (i, j, k, l: integer; var t: perm);
  var
    m: integer;
begin
  t.size := i + j + k + l;
  for m := 1 to i + j + k + l do
    if m <= i then
      t.data[m] := m
    else if m <= i + j then
      t.data[m] := m + k
    else if m <= i + j + k then
```

```
            t.data[m] := m — j
         else
            t.data[m] := m;
end; {vperm}
procedure copyperm (p1: perm; var p2: perm);
   var
      i: integer;
begin
   p2.size := p1.size;
   for i := 1 to p1.size do
      p2.data[i] := p1.data[i];
end; {copyperm}
procedure permult (p1, p2: perm; var outp: perm);
   var
      i: integer;
begin
   if p1.size = 0 then
      copyperm(p2, outp)
   else if p2.size = 0 then
      copyperm(p1, outp)
   else
      begin
         outp.size := max(p1.size, p2.size);
         for i := 1 to outp.size do
            outp.data[i] := i;
         for i := 1 to outp.size do
            begin
               if i <= p2.size then
                  outp.data[i] := p2.data[i];
               if outp.data[i] <= p1.size then
                  outp.data[i] := p1.data[outp.data[i]];
            end
      end
end; {permult}
function invper (m: perm; inp: integer): integer;
   var
      i, outp: integer;
begin
   outp := —1;
   if (m.size = 0) or (inp > m.size) then
      outp := inp
   else
      for i := 1 to m.size do
         if m.data[i] = inp then
            outp := i;
   if outp = —1 then
      writeln('Original input was not a permutation.')
   else
      invper := outp;
end; {invper}
function idperm (a: perm): boolean;
   var
      s: boolean;
      i: integer;
begin
```

```
    s := true;
    if a.size > 0 then
       for i := 1 to a.size do
          s := (s and (a.data[i] = i));
    idperm := s;
end; {idperm}
procedure tfunct_simple (var m: perm; pivot: integer);
    var
       i, beta: integer;
begin
    if pivot <= m.size then
       begin
          beta := invper(m, pivot);
          for i := 1 to m.size do
             if m.data[i] > pivot then
                m.data[i] := m.data[i] + 1;
          for i := m.size downto beta + 1 do
             m.data[i + 1] := m.data[i];
          m.data[beta + 1] := pivot + 1;
          m.size := m.size + 1;
       end;
end;{tfunct_simple}
procedure concatseq (var outs: numseq; var in1, in2: numseq)
    var
       i;
begin
    outs.size := in1.size + in2.size;
    for i := 1 to in1.size do
       outs.data[i] := in1.data[i];
    for i := 1 to in2.size do
       outs.data[i + in1.size − 1] := in2.data[i];
end;
procedure tfunct (var inp: perm; var outp: perm; var pivot: numseq);
    var
       lseq: array[1..psize] of numseq;
begin
end;{tfunct}
procedure tfrakturterm (inp: barptr; pivot: integer; var outp: barptr);
begin
    if inp = nil then
       outp := nil
    else
       begin
          new(outp);
          copyperm(inp^.first, outp^.first);
          tfunct_simple(outp^.first, pivot);
          tfrakturterm(inp^.rest, invper(inp^.first, pivot), outp^.rest)
       end;
end;
procedure tfraktur (inp: bardatptr; pivot: integer; var outp: bardatptr);
begin
    if inp = nil then
       outp := nil
    else
       begin
          new(outp);
```

```
              with outp^.coef do
                 begin
                    numval := inp^.coef.numval;
                    grpelt := inp^.coef.grpelt;
                    tfunct_simple(grpelt, pivot);
                 end;
              tfrakturterm(inp^.term, invper(inp^.coef.grpelt, pivot), outp^.term);
              tfraktur(inp^.rest, pivot, outp^.rest);
           end;
     end;
procedure dispper (m: perm);
   type
      eset = set of 1..psize;
   var
      dset: eset;
      cmat: permat;
      lowelt, i, csize, celt: integer;
      printsw: boolean;
begin
   if idperm(m) then
      write('Identity')
   else
      begin
         printsw := false;
         dset := [1..m.size];
         while dset <> [] do
            begin
               lowelt := 0;
               repeat
                  lowelt := lowelt + 1;
               until lowelt in dset;
               csize := 1;
               celt := lowelt;
               repeat
                  cmat[csize] := celt;
                  dset := dset − [celt];
                  celt := m.data[celt];
                  csize := csize + 1;
               until celt = lowelt;
               csize := csize − 1;
               if csize > 1 then
                  begin
                     printsw := true;
                     write(' (');
                     for i := 1 to csize − 1 do
                        begin
                           putnum(cmat[i]);
                           write(',')
                        end;
                     putnum(cmat[csize]);
                     write(') ');
                  end;
            end;
         if not printsw then
            write('Identity');
      end
```

```
  end; {dispper}
  procedure putbareltptr (inp: barptr);
    procedure inner (inp: barptr);
    begin
      if inp <> nil then
        with inp^ do
          begin
            dispper(first);
            if rest <> nil then
              begin
                write(' | ');
                inner(rest);
              end
          end;
    end;
  begin
    write(' [ ');
    inner(inp);
    write(' ] ');
  end; {putbareltptr}
  procedure putcoef (c: coeftype);
  begin
    if not (c.numval in [-1, 1]) then
      putnum(c.numval);
    if c.numval = -1 then
      write(' - ');
    if not idperm(c.grpelt) then
      dispper(c.grpelt)
  end; {putcoef}
  procedure putbardatptr (inp: bardatptr);
  begin
    if inp <> nil then
      with inp^ do
        begin
          if coef.numval <> 0 then
            begin
              putcoef(coef);
              putbareltptr(term);
              writeln;
            end;
          putbardatptr(rest)
        end;
  end; {putbardatptr}
  function eqperm (a, b: perm): boolean;
    var
      i: integer;
      c: boolean;
  begin
    c := true;
    if (a.size = 0) and (b.size > 0) then
      for i := 1 to a.size do
        c := c and (b.data[i] = i);
    if (a.size > 0) and (b.size = 0) then
      for i := 1 to a.size do
        c := c and (a.data[i] = i);
    if (a.size > 0) and (b.size > 0) then
```

```
       if a.size = b.size then
          for i := 1 to a.size do
             c := c and (a.data[i] = b.data[i])
       else
          c := false;
    eqperm := c;
end; {eqperm}
function copybarelt (a: barptr): barptr;
    var
       c: barptr;
begin
   if a = nil then
      copybarelt := nil
   else
      begin
         new(c);
         copyperm(a^.first, c^.first);
         c^.rest := copybarelt(a^.rest);
         copybarelt := c
      end;
end; {copybarelt}
procedure multcoefbardat (a: coeftype; b: bardatptr; var outp: bardatptr);
    var
       temp1: bardatptr;
begin
   if a.numval = 0 then
      outp := nil
   else if b = nil then
      outp := nil
   else
      begin
         new(temp1);
         temp1^.coef.numval := a.numval * b^.coef.numval;
         permult(a.grpelt, b^.coef.grpelt, temp1^.coef.grpelt);
         temp1^.term := copybarelt(b^.term);
         multcoefbardat(a, b^.rest, temp1^.rest);
         outp := temp1;
      end
end; {multcoefbardat}
function eqbareltptr (a, b: barptr): boolean;
    var
       c: boolean;
begin
   if a = nil then
      c := (b = nil)
   else if b = nil then
      c := false
   else
      c := (eqperm(a^.first, b^.first) and eqbareltptr(a^.rest, b^.rest));
   eqbareltptr := c;
end; {eqbareltptr}
function dim (a: barptr): integer;
begin
   if a = nil then
      dim := 0
```

```
      else
         dim := dim(a^.rest) + 1
end; {dim}
function addonebareltptr (a: bardatptr; b: barptr; c: coeftype): bardatptr;
   var
      d: bardatptr;
begin
   if a = nil then
      begin
         new(d);
         d^.coef.numval := c.numval;
         copyperm(c.grpelt, d^.coef.grpelt);
         d^.term := copybarelt(b);
         d^.rest := nil
      end
   else if (eqbareltptr(a^.term, b) and eqperm(a^.coef.grpelt, c.grpelt)) then
      begin
         a^.coef.numval := a^.coef.numval + c.numval;
         if a^.coef.numval = 0 then
            begin
               d := a^.rest;
            end
         else
            d := a
      end
   else
      begin
         d := a;
         a^.rest := addonebareltptr(a^.rest, b, c);
      end;
   addonebareltptr := d;
end; {addonebareltptr}
function sum (a, b: bardatptr): bardatptr;
begin
   if a = nil then
      sum := b
   else if b = nil then
      sum := a
   else
      begin
         sum := addonebareltptr(sum(a, b^.rest), b^.term, b^.coef)
      end
end; {sum}
procedure killbarelt (a: barptr);
begin
   if a <> nil then
      if a^.rest <> nil then
         begin
            killbarelt(a^.rest);
            dispose(a)
         end
      else
         dispose(a)
end; {killbarelt}
procedure killbardatptr (a: bardatptr);
```

```
begin
  if a <> nil then
    if a^.rest <> nil then
      begin
        killbardatptr(a^.rest);
        killbarelt(a^.term);
        dispose(a)
      end
    else
      begin
        killbarelt(a^.term);
        dispose(a)
      end
end; {killbardatptr}
function shuffleone (a, b: barptr): bardatptr;
  type
    smat = array[1..20] of integer;
  var
    v, w: array[1..20] of barptr;
    i: integer;
    sign: coeftype;
    outstuff: bardatptr;
    totdim, dima, dimb: integer;
    s: smat;
  procedure putshuffle;
    var
      i, acount, bcount: integer;
      lastz, zhold: barptr;
  begin
    acount := dima + 1;
    bcount := dimb + 1;
    for i := totdim downto 1 do
      begin
        case s[i] of
          1:
            begin
              acount := acount - 1;
              if i < totdim then
                lastz^.rest := v[acount];
              lastz := v[acount];
              if i < totdim then
                lastz^.rest := nil;
            end;
          0:
            begin
              bcount := bcount - 1;
              if i < totdim then
                lastz^.rest := w[bcount];
              lastz := w[bcount];
              if i < totdim then
                lastz^.rest := nil;
            end;
        end;
        if i = totdim then
          zhold := lastz;
      end;
```

```
      outstuff := addonebareltptr(outstuff, zhold, sign);
  end; {putshuffle}
  procedure genshuf (a, b, start: integer);
    var
      i, j, k: integer;
  begin
    if a = 0 then
      begin
        for i := start to start − 1 + b do
          s[i] := 0;
        putshuffle;
      end
    else if b = 0 then
      begin
        for i := start to start − 1 + a do
          s[i] := 1;
        putshuffle;
      end
    else
      for i := start to start + b do
        begin
          k := 0;
          for j := start to i − 1 do
            begin
              sign.numval := −sign.numval;
              k := k + 1;
              s[j] := 0
            end;
          s[i] := 1;
          genshuf(a − 1, b − k, i + 1)
        end;
  end;{genshuf}
begin
  if a = nil then
    begin
      new(outstuff);
      with outstuff^ do
        begin
          term := copybarelt(b);
          coef.numval := 1;
          coef.grpelt.size := 0;
          rest := nil
        end;
      shuffleone := outstuff
    end
  else if b = nil then
    begin
      new(outstuff);
      with outstuff^ do
        begin
          term := copybarelt(a);
          coef.numval := 1;
          coef.grpelt.size := 0;
          rest := nil
        end;
      shuffleone := outstuff
```

```
      end
    else
      begin
        dima := dim(a);
        dimb := dim(b);
        sign.numval := 1;
        sign.grpelt.size := 0;
        totdim := dima + dimb;
        for i := dima downto 1 do
          begin
            v[i] := copybarelt(a);
            v[i]^.rest := nil;
            a := a^.rest;
          end;
        for i := dimb downto 1 do
          begin
            w[i] := copybarelt(b);
            w[i]^.rest := nil;
            b := b^.rest;
          end;
        outstuff := nil;
        genshuf(dima, dimb, i );
        shuffleone := outstuff
      end
end; {shuffleone}
function shiftone (a: barptr; amount: integer): barptr;
  var
    z: barptr;
    i: integer;
begin
  if a <> nil then
    begin
      new(z);
      z := copybarelt(a);
      z^.first.size := z^.first.size + amount;
      for i := z^.first.size downto amount + 1 do
        z^.first.data[i] := z^.first.data[i − amount] + amount;
      for i := 1 to amount do
        z^.first.data[i] := i;
      z^.rest := shiftone(a^.rest, amount);
      shiftone := z
    end
  else
    shiftone := nil
end; {shiftone}
function shift (a: bardatptr; amount: integer): bardatptr;
  var
    z: bardatptr;
    i: integer;
begin
  if a <> nil then
    begin
      new(z);
      z^.term := shiftone(a^.term, amount);
      z^.coef.numval := a^.coef.numval;
      z^.coef.grpelt.size := 0;
```

```
        if a^.coef.grpelt.size > 0 then
            begin
                z^.coef.grpelt.size := a^.coef.grpelt.size + amount;
                for i := z^.coef.grpelt.size downto amount + 1 do
                    z^.coef.grpelt.data[i] := a^.coef.grpelt.data[i − amount] + amount;
                for i := 1 to amount do
                    z^.coef.grpelt.data[i] := i;
            end;
            z^.rest := shift(a^.rest, amount);
            shift := z
        end
    else
        shift := nil;
end; {shift}
function shuffle (a, b: bardatptr): bardatptr;
    var
        z, g: bardatptr;
        r, s: barptr;
        temp: coeftype;
    procedure scalarmult (a: bardatptr; b: coeftype);
    begin
        if a <> nil then
            begin
                a^.coef.numval := a^.coef.numval * b.numval;
                permult(b.grpelt, a^.coef.grpelt, a^.coef.grpelt);
                scalarmult(a^.rest, b);
            end
    end;
begin
    if b = nil then
        shuffle := nil
    else if a = nil then
        shuffle := nil
    else
        begin
            r := a^.term;
            s := b^.term;
            z := shuffleone(r, s);
            permult(a^.coef.grpelt, b^.coef.grpelt, temp.grpelt);
            temp.numval := a^.coef.numval * b^.coef.numval;
            scalarmult(z, temp);
            new(g);
            g^.coef := a^.coef;
            g^.rest := nil;
            g^.term := copybarelt(a^.term);
            z := sum(z, shuffle(g, b^.rest));
            killbarelt(g^.term);
            g^.coef := b^.coef;
            g^.rest := nil;
            g^.term := copybarelt(b^.term);
            z := sum(z, shuffle(a^.rest, g));
            killbardatptr(g);
            z := sum(z, shuffle(a^.rest, b^.rest));
            shuffle := z
        end
end; {shuffle}
```

```
function phi (a: bardatptr): bardatptr;
   var
      z: bardatptr;
      w: barptr;
begin
   if a = nil then
      phi := nil
   else
      begin
         if idperm(a^.coef.grpelt) then
            z := phi(a^.rest)
         else
            begin
               new(z);
               new(w);
               z^.coef.numval := a^.coef.numval;
               z^.coef.grpelt.size := 0;
               w^.first := a^.coef.grpelt;
               w^.rest := a^.term;
               z^.term := w;
               z^.rest := phi(a^.rest);
            end;
         phi := z
      end;
end; {phi}
function badterm (inp: barptr): boolean;
   var
      tarray1, tarray2: array[1..psize] of perm;
      curr: barptr;
      count, i, j: integer;
      testset: set of 1..psize;
begin
   curr := inp;
   count := 1;
   while (curr <> nil) do
      begin
         tarray1[count] := curr^.first;
         curr := curr^.rest;
         count := count + 1;
      end;
   count := count − 1;
   for i := 1 to count do
      begin
         tarray2[i] := tarray1[i];
         for j := i + 1 to count do
            permult(tarray2[i], tarray1[j], tarray2[i]);
      end;
   testset := [1..count];
   for i := 1 to count do
      testset := testset − [invper(tarray2[i], count + 1)];
   badterm := (testset = []);
end; {testterm}
procedure testresult (inp: bardatptr; i, j: integer);
begin
   if inp <> nil then
      begin
```

```
        if badterm(inp^.term) and not ((i = 1) and (j = 1)) then
          begin
            write('The following term  gives a');
            write('nontrivial result when evaluated on the ');
            write('m-structure of the bar contruction');
            writeln('-- a compensating boundary must be added:');
            putbareltptr(inp^.term)
          end;
        testresult(inp^.rest, i, j);
      end;
end;
function genval (u, v: integer): bardatptr;
  var
    accum, t1, t2, t3, t4, t5, t: bardatptr;
    i, j: integer;
    genterm: barptr;
    signcomp: coeftype;
  procedure gen (i, j: integer);
  begin
    vperm(i, j, u − i, v − j, signcomp.grpelt);
    signcomp.numval := vsign(j, u − i);
    t1 := genval(i, j);
    t2 := genval(u − i, v − j);
    t3 := shift(t2, i + j);
    t4 := shuffle(t1, t3);
    if odd((i + j − 1) * (u − i + v − j) + u + v) then
      signcomp.numval := −signcomp.numval;
    multcoefbardat(signcomp, t4, t5);
    t := phi(t5);
    accum := sum(accum, t);
    killbardatptr(t1);
    killbardatptr(t2);
    killbardatptr(t3);
    killbardatptr(t4);
  end; {gen}
begin
  if ((u = 1) and (v = 1) and (mode = canonical)) then
    begin
      new(accum);
      with accum^ do
        begin
          with coef do
            begin
              numval := 1;
              with grpelt do
                begin
                  size := 2;
                  data[1] := 2;
                  data[2] := 1
                end;
            end;
          rest := nil
        end;
      new(genterm);
      with genterm^ do
        begin
```

```
                first.size := 2;
                first.data[1] := 2;
                first.data[2] := 1;
                rest := nil
              end;
          accum^.term := genterm;
          genval := accum;
        end
      else if (((u = 0) and (v = 1)) or ((u = 1) and (v = 0))) then
        begin
          new(accum);
          with accum^ do
            begin
              coef.numval := 1;
              coef.grpelt.size := 0;
              rest := nil;
              term := nil
            end;
          genval := accum;
        end
      else if ((u = 0) and (v > 1)) or ((u > 1) and (v = 0)) or ((u = 0) and (v = 0)) then
        genval := nil
      else
        begin
          accum := nil;
          for i := 0 to u do
            for j := 0 to v do
              if (((i > 0) or (j > 0)) and ((i < u) or (j < v))) then
                gen(i, j);
          if (u > 1) and (mode <> simple) then
            for i := 1 to u − 1 do
              begin
                tfraktur(genval(u − 1, v), i, t1);
                if odd(u + v + i) then
                  signcomp.numval := −1
                else
                  signcomp.numval := 1;
                signcomp.grpelt.size := 0;
                multcoefbardat(signcomp, t1, t2);
                t3 := phi(t2);
                accum := sum(accum, t3);
                killbardatptr(t1);
                killbardatptr(t2);
              end;
          if (v > 1) and (mode <> simple) then
            for i := u + 1 to u + v − 1 do
              begin
                tfraktur(genval(u, v − 1), i, t1);
                if odd(u + v + i) then
                  signcomp.numval := −1
                else
                  signcomp.numval := 1;
                signcomp.grpelt.size := 0;
                multcoefbardat(signcomp, t1, t2);
                t3 := phi(t2);
                accum := sum(accum, t3);
```

```
                    killbardatptr(t1);
                    killbardatptr(t2);
                end;
            genval := accum;
        end;
    end; {genval}
begin {main program}
    repeat
        writeln('Enter mode:  s for simple or c for canonical.');
        readln(resp);
        if resp in ['s', 'S'] then
            mode := simple
        else
            mode := canonical;
        writeln('Enter two subscripts:');
        readln(i, j);
        t1 := genval(i, j);
        putbardatptr(t1);
        if mode = canonical then
            testresult(t1, i, j);
        killbardatptr(t1);
        writeln('Do you want to compute another s-function value?');
        readln(resp);
    until not (resp in ['Y', 'y']);
end.
```

Bibliography

[1] J. Adams, On the cobar construction, 81–87, Georges Thone, Liège and Masson, Paris, 1957, pp. 81–87.

[2] Hans Baues, *Geometry of loop spaces and the cobar construction*, (Providence, Rhode Island, USA), vol. 25, Memoirs of the A. M. S., no. 230, Providence, Rhode Island, USA, 1980.

[3] E. Brown, *Twisted tensor products, I*, Ann. of Math. (2) **69** (1959), 223–246.

[4] Ron Brown, *The twisted Eilenberg-Zilber theorem*, Celebrazioni archimedee del secolo XX Simposio de topologia (1964), 33–37.

[5] H. Cartan, *Algebras d'Eilenberg-MacLane et homotopie*, Seminaire Henri Cartan (1954/55).

[6] James F. Davis, *Higher diagonal approximations and skeletons of $K(\pi, 1)$'s*, Lecture Notes in Mathematics, vol. 1126, Springer-Verlag, 1983, pp. 51–61.

[7] Samuel Eilenberg and Saunders MacLane, *On the groups $H(\Pi, n)$. I*, Ann. of Math. (2) **58** (1954), 55–106.

[8] ———, *On the groups $H(\Pi, n)$. II*, Ann. of Math. (2) **60** (1954), 49–139.

[9] V. K. A. M. Gugenheim, *On a theorem of E. H. Brown*, Illinois J. Math. **4** (1960), 292–311.

[10] ———, *On the chain-complex of a fibration*, Illinois J. Math. **16** (1972), 398–414.

[11] V.K.A.M. Gugenheim, M. G. Barrett, and J. C. Moore, *On semisimplicial fiber-bundles*, Amer. J. Math. **81** (1959), 639–657.

[12] Stephen Halperin and James Stesheff, *Obstructions to homotopy equivalences*, Adv. in Math. **32** (1979), 233–279.

[13] P. J. Hilton and S. Wylie, *Homology theory*, Cambridge University Press, 1965.

[14] Tornike V. Kadeishvili, *О категории дифференциальних коалгебр и категорий $A(\infty)$-алгебр*, Trudy Tbiliss. Mat. Inst. Razmadze Akad. Nauk Gruzin. SSR **77** (1985), 51–70.

[15] J. P. May, *The geometry of iterated loop-spaces*, Lecture Notes in Mathematics, vol. 271, Springer-Verlag, 1972.

[16] R. James Milgram, *The bar construction and abelian H-spaces*, Illinois J. Math. **11** (1967), 242–250.

[17] A. Prouté, *A_∞-structures, modèle minimal de Baues-Lemaire et homologie des fibrations*, preprint.

[18] ———, *Sur la transformation d'Eilenberg-MacLane*, C. R. Acad. Sci. Paris Sér. I Math. **297** (1983), 193–194.

[19] ———, *Un contre-exemple à la géométricité du shuffle-coproduit de la cobar-construction*, C. R. Acad. Sci. Paris Sér. I Math. **298** (1984), no. 2, 31–34.

[20] D. Quillen, *Rational homotopy theory*, Ann. of Math. (2) **90** (1969), 205–295.

[21] Samson Saneblidze, *Filtered model of a fibration and rational obstruction theory*, Manuscripta Math. **76** (1992), 111–136.

[22] V. A. Smirnov, *Homology of fiber spaces*, Uspekhi Mat. Nauk **35** (1980), no. 3, 183–188.

[23] ———, *Homotopy theory of coalgebras*, Izv. Akad. Nauk SSSR Ser. Mat. **49** (1985), no. 6, 575–592.

[24] Justin Smith, *Algebraic homotopy: I. m-coalgebras*, preprint.

[25] ———, *Equivariant Moore spaces*, (Berlin, Heidelberg, New York, Tokyo) (Andrew Ranicki, Norman Levitt, and Frank Quinn, eds.), Lecture Notes in Mathematics, vol. 1126, Springer-Verlag, Berlin, Heidelberg, New York, Tokyo, 1983, pp. 238–270.

[26] ———, *Equivariant Moore spaces. II — The low-dimensional case*, J. Pure Appl. Algebra **36** (1985), 187–204.

[27] James D. Stasheff, *Homotopy associativity of h-spaces I, II*, Trans. Amer. Math. Soc. **108** (1963), 275–312.

[28] R. H. Szczarba, *The homology of twisted cartesian products*, Trans. Amer. Math. Soc. **100** (1961), 197–216.

[29] D. Tanré, *Homotopie rationelle: modèles de Chen, Quillen, Sullivan*, Lecture Notes in Mathematics **1025** (1982).

[30] В. А. Смирнов, *Когомологии алгебры Стинрода*, Математические заметки **52** (1992), no. 2, 120–125.

[31] _____, *Вторичные операции в гомологиях операды E*, Известия Ак. Наук — Серия Математическая **56** (1992), no. 2, 449–468.

[32] Shih Weishu, *Homologie des espaces fibrés*, Inst. Hautes Études Sci. Publ. Math. **13** (1962), 93–176.

Index

DEPARTMENT OF MATHEMATICS AND COMPUTER SCIENCE, DREXEL UNIVERSITY, PHILADELPHIA, PA 19104
E-mail address: jsmith@mcs.drexel.edu

Editorial Information

To be published in the *Memoirs*, a paper must be correct, new, nontrivial, and significant. Further, it must be well written and of interest to a substantial number of mathematicians. Piecemeal results, such as an inconclusive step toward an unproved major theorem or a minor variation on a known result, are in general not acceptable for publication. *Transactions* Editors shall solicit and encourage publication of worthy papers. Papers appearing in *Memoirs* are generally longer than those appearing in *Transactions* with which it shares an editorial committee.

As of March 4, 1994, the backlog for this journal was approximately 7 volumes. This estimate is the result of dividing the number of manuscripts for this journal in the Providence office that have not yet gone to the printer on the above date by the average number of monographs per volume over the previous twelve months, reduced by the number of issues published in four months (the time necessary for preparing an issue for the printer). (There are 6 volumes per year, each containing at least 4 numbers.)

A Copyright Transfer Agreement is required before a paper will be published in this journal. By submitting a paper to this journal, authors certify that the manuscript has not been submitted to nor is it under consideration for publication by another journal, conference proceedings, or similar publication.

Information for Authors and Editors

Memoirs are printed by photo-offset from camera copy fully prepared by the author. This means that the finished book will look exactly like the copy submitted.

The paper must contain a *descriptive title* and an *abstract* that summarizes the article in language suitable for workers in the general field (algebra, analysis, etc.). The *descriptive title* should be short, but informative; useless or vague phrases such as "some remarks about" or "concerning" should be avoided. The *abstract* should be at least one complete sentence, and at most 300 words. Included with the footnotes to the paper, there should be the 1991 *Mathematics Subject Classification* representing the primary and secondary subjects of the article. This may be followed by a list of *key words and phrases* describing the subject matter of the article and taken from it. A list of the numbers may be found in the annual index of *Mathematical Reviews*, published with the December issue starting in 1990, as well as from the electronic service e-MATH [**telnet e-MATH.ams.org** (or **telnet 130.44.1.100**). Login and password are **e-math**]. For journal abbreviations used in bibliographies, see the list of serials in the latest *Mathematical Reviews* annual index. When the manuscript is submitted, authors should supply the editor with electronic addresses if available. These will be printed after the postal address at the end of each article.

Electronically prepared manuscripts. The AMS encourages submission of electronically prepared manuscripts in $\mathcal{AMS}$-TEX or $\mathcal{AMS}$-LATEX because properly prepared electronic manuscripts save the author proofreading time and move more quickly through the production process. To this end, the Society has prepared "preprint" style files, specifically the amsppt style of $\mathcal{AMS}$-TEX and the amsart style of $\mathcal{AMS}$-LATEX, which will simplify the work of authors and of the production staff. Those authors who make use of these style files from the beginning of the writing process will further reduce their own effort. Electronically submitted manuscripts prepared in plain $T_{E}X$ or $L^{A}T_{E}X$ do not mesh properly with the AMS production systems and cannot, therefore, realize the same kind of expedited processing. Users of plain $T_{E}X$ should have little difficulty learning $\mathcal{A}_{\mathcal{M}}\mathcal{S}$-$T_{E}X$, and $L^{A}T_{E}X$ users will find that $\mathcal{A}_{\mathcal{M}}\mathcal{S}$-$L^{A}T_{E}X$ is the same as $L^{A}T_{E}X$ with additional commands to simplify the typesetting of mathematics.

Guidelines for Preparing Electronic Manuscripts provides additional assistance and is available for use with either $\mathcal{A}_{\mathcal{M}}\mathcal{S}$-$T_{E}X$ or $\mathcal{A}_{\mathcal{M}}\mathcal{S}$-$L^{A}T_{E}X$. Authors with FTP access may obtain *Guidelines* from the Society's Internet node e-MATH.ams.org (130.44.1.100). For those without FTP access *Guidelines* can be obtained free of charge from the e-mail address guide-elec@ math.ams.org (Internet) or from the Customer Services Department, American Mathematical Society, P.O. Box 6248, Providence, RI 02940-6248. When requesting *Guidelines*, please specify which version you want.

At the time of submission, authors should indicate if the paper has been prepared using $\mathcal{A}_{\mathcal{M}}\mathcal{S}$-$T_{E}X$ or $\mathcal{A}_{\mathcal{M}}\mathcal{S}$-$L^{A}T_{E}X$. The *Manual for Authors of Mathematical Papers* should be consulted for symbols and style conventions. The *Manual* may be obtained free of charge from the e-mail address cust-serv@math.ams.org or from the Customer Services Department, American Mathematical Society, P.O. Box 6248, Providence, RI 02940-6248. The Providence office should be supplied with a manuscript that corresponds to the electronic file being submitted.

Electronic manuscripts should be sent to the Providence office immediately after the paper has been accepted for publication. They can be sent via e-mail to pub-submit@math.ams.org (Internet) or on diskettes to the Publications Department, American Mathematical Society, P.O. Box 6248, Providence, RI 02940-6248. When submitting electronic manuscripts please be sure to include a message indicating in which publication the paper has been accepted.

Two copies of the paper should be sent directly to the appropriate Editor and the author should keep one copy. The *Guide for Authors of Memoirs* gives detailed information on preparing papers for *Memoirs* and may be obtained free of charge from the Editorial Department, American Mathematical Society, P.O. Box 6248, Providence, RI 02940-6248. For papers not prepared electronically, model paper may also be obtained free of charge from the Editorial Department.

Any inquiries concerning a paper that has been accepted for publication should be sent directly to the Editorial Department, American Mathematical Society, P.O. Box 6248, Providence, RI 02940-6248.